Ray, Geberekrstos, Orasugh, Muzata

Process-Induced Phase Separation in Polymer Blends

Suprakas Sinha Ray
Amanuel Geberekrstos
Jonathan Tersur Orasugh
Tanyaradzwa Sympathy Muzata

Process-Induced Phase Separation in Polymer Blends

Materials, Characterization, Properties, and Applications

HANSER
Hanser Publishers, Munich

The Authors:

Suprakas Sinha Ray, Centre for Nanostructures and Advanced Materials, DSI-CSIR Nanotechnology Innovation Centre, Council for Scientific and Industrial Research, Pretoria, South Africa; Department of Chemical Sciences, University of Johannesburg, Doornfontein, Johannesburg, South Africa, ORCID: 0000-0002-0007-2595

Amanuel Geberekrstos, Department of Chemical Sciences, University of Johannesburg, Doornfontein, Johannesburg, South Africa; Centre for Nanostructures and Advanced Materials, DSI-CSIR Nanotechnology Innovation Centre, Council for Scientific and Industrial Research, Pretoria, South Africa

Jonathan Tersur Orasugh, Department of Chemical Sciences, University of Johannesburg, Doornfontein, Johannesburg, South Africa; Centre for Nanostructures and Advanced Materials, DSI-CSIR Nanotechnology Innovation Centre, Council for Scientific and Industrial Research, Pretoria, South Africa

Tanyaradzwa Sympathy Muzata, Department of Polymer Technology and Engineering, Harare Institute of Technology, Belvedere, Harare, Zimbabwe; Currently working as a postdoctoral Researcher, School of Packaging, Michigan State University, USA

FSC
www.fsc.org
MIX
Paper from responsible sources
FSC® C083411

Distributed by:
Carl Hanser Verlag
Postfach 86 04 20, 81631 Munich, Germany
Fax: +49 (89) 98 48 09
www.hanserpublications.com
www.hanser-fachbuch.de

Editor: Mark Smith
Production Management: Cornelia Speckmaier
Cover design: Max Kostopoulos
Cover picture: Suprakas Sinha Ray
Typesetting: Eberl & Koesel Studio, Kempten, Germany
Printed and bound by CPI books GmbH, Leck
Printed in Germany

ISBN: 978-1-56990-901-0
E-Book ISBN: 978-1-56990-919-5

Progress in Polymer Processing (PPP) Series

Satinath Bhattacharya, *Series Editor*
RMIT University, Australia

Editorial Advisory Board

Ica **Manas-Zloczower**
Case Western Reserve University
U.S.A.

Masami **Okamoto**
Toyota Technological Institute
Japan

Chul **Park**
University of Toronto
Canada

Luiz A. **Pessan**
Universidade Federal de São Carlos
Brazil

Changyu **Shen**
Dalian University of Technology
China

Mark **Smith**
Carl Hanser Verlag GmbH & Co. KG
Germany

Giuseppe **Titomanlio**
Università degli Studi di Salerno
Italy

Lih-Sheng (Tom) **Turng**
University of Wisconsin
U.S.A.

John **Vlachopoulos**
McMaster University
Canada

The Authors

Suprakas Sinha Ray is a Chief Research Scientist and Manager of the Centre for Nanostructures and Advanced Materials, DSI-CSIR Nanotechnology Innovation Centre, Council for Scientific and Industrial Research, Pretoria, South Africa. He received his PhD degree in Physical Chemistry at the University of Calcutta, India in 2001 and was a recipient of the "Sir P. C. Ray Research Award" for the best PhD work. Prof. Ray's current research focuses on the applications of advanced nanostructured and polymeric materials. He is one of the most active and highly cited authors in the field of polymer nanocomposite materials, and he has recently been rated by Thomson Reuters as being one of the top 1% most impactful and influential scientists and top 50 high-impact chemists.

Prof. Ray is the author of five books, co-author of five edited books and 32 book chapters on various aspects of polymer-based nanostructured materials and their applications, and author and co-author of 455 articles in high-impact international journals and 30 articles in national and international conference proceedings. He also has six patents and seven newly demonstrated technologies shared with colleagues, collaborators and industrial partners to his name. So far, his team commercialized 19 different products. His prestigious honors and awards include South Africa's 2016 National Science and Technology Award (NSTF); 2014 CSIR-wide Leadership Award; 2014 CSIR Human Capital Development Award; and 2013 Morand Lambla Award (top award in the field of polymer processing worldwide), International Polymer Processing Society, USA. He is also appointed as Extraordinary Professor at the University of Pretoria and Distinguished Professor of Chemistry, University of Johannesburg.

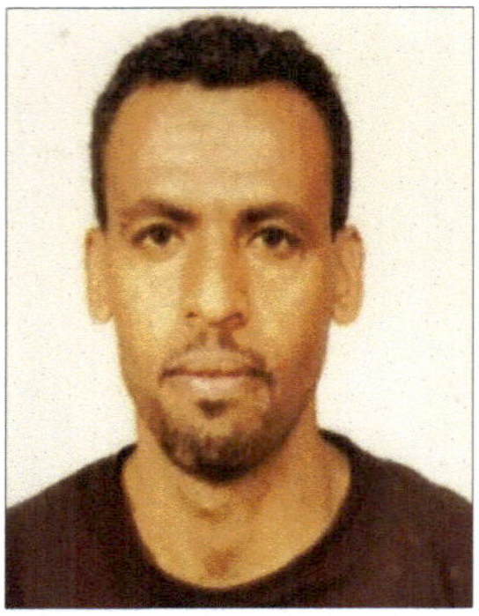

Amanuel Geberekrstos received his PhD in Chemical Engineering from the Indian Institute of Science, Bangalore, India. Currently, he is working as a Postdoctoral Fellow at the Department of Chemical Sciences, University of Johannesburg and also associated as a senior researcher at the Centre for Nanostructures and Advanced Materials, DSI-CSIR Nanotechnology Innovation Centre, Council for Scientific and Industrial Research, Pretoria, South Africa. His research work focuses on the effects of nanoparticle localization and geometry on the morphology development, and electrical and dielectric properties of immiscible blends, in association with their rheological properties. He is the author of 20 articles in international journals and filed one patent on various aspects of phase-separated polymer blends and their application in advanced membrane and EMI shielding applications.

Jonathan Tersur Orasugh received his PhD in Polymer Science and Engineering from the University of Calcutta, India. Currently, he is working as a Postdoctoral Fellow at the Department of Chemical Sciences, University of Johannesburg and also associated as a senior researcher at Centre for Nanostructures and Advanced Materials, DSI-CSIR Nanotechnology Innovation Centre, Council for Scientific and Industrial Research, Pretoria, South Africa. His research work focuses on the processing and characterization of biopolymer based immiscible polymer blends for biomedical applications, particularly drug delivery and tissue engineering.

Tanyaradzwa Sympathy Muzata received his PhD in Materials Engineering from the Indian Institute of Science, Bangalore, India. He is working as a Lecturer at Department of Polymer Technology and Engineering, Harare Institute of Technology, Belvedere, Harare, Zimbabwe. Currently, he is undertaking postdoctoral study at School of Packaging, Michigan State University, USA. His research work focuses on understanding the demixing behavior and structure-property correlation in graphene oxide containing LCST blends. He is the authors of several publications on nanoparticles containing polymer blends and their applications in energy storage and EMI shielding.

Preface

Phase separation in polymer blends has attracted tremendous techno-commercial importance from fundamental to practical applications. Most applications of polymer blends, such as tissue engineering, membrane technology, electromagnetic shielding, energy harvesting, structural materials, packaging, smart multiphase polymer coatings, and others, depend on the morphologies developed during processing. Conventionally, phase-separated polymer blends can be prepared and their properties tuned via various processing techniques. However, in some cases, their processing methods using predetermined conventional polymer processing techniques require modifications before usage. The dynamic properties of these polymers are strongly linked with stress-diffusion coupling and viscoelastic effects. It is, in theory, possible to examine the nucleation-limited mechanism of phase separation in polymer blends; nevertheless, experimentally, it is more challenging. Diverse phase-separated polymeric structures can be prepared depending on the thermodynamic aspects of the polymer pairs, where it is important to understand the basics of the phase separations in these polymer blends within the context of the miscibility of these polymers. In the research and/or review literature reported to date it is shown that several properties such as morphologies, response to external stimuli, and mechanical properties of phase-separated polymers as per intended application can be created by simply altering the dynamic and kinetic parameters such as the viscosity ratio, blend ratio, mixing time, compatibilization, and mixer sort.

Currently, there is no dedicated authored book on this topic; the few review articles and reports either concentrate on specific phase-separated polymer blends and perspectives or deliver only short descriptions with limited instances. Thus, there is a need for a comprehensive, authored book presenting an exhaustive and serious overview of state of the art in phase-separated or multiphasic polymer blends, including significant past and recent developments. Moreover, there is a need to address fundamental issues that persist regarding phase-separated polymer blends and their nanocomposites, which could profile future research and industrial application of these materials for a novel sustainable future development. This book emphasizes recent research developments, processing techniques, characterization methods, factors influencing phase separation temperature in phase-separated, including partially miscible, polymer blends, and key research challenges in developing phase-separated polymer materials as per literature in different application areas.

The field of polymer blends has rapidly grown over the past years, as witnessed by the number of publications being produced yearly. Since polymer blends are a vast area, it is of paramount importance to fully exploit it from a fundamental aspect to applying the multiphasic polymeric material. The journey from fundamentals to the application of polymer blends is one area that is still lacking, and this book is systematically structured to cater to this need. This book will extensively cover the thermodynamics of polymer blends and the fabrication of multiphasic polymeric systems with a precise and unique architecture.

This book will be most valuable to the reader as it will extensively outline (i) the fundamental aspect of polymer blend thermodynamics, (ii) the state-of-the-art processing techniques for specific polymer blend systems currently in use, and (iii) the design and fabrication of multiphasic polymeric materials, which will present a multiplicity of opportunities in water remediation, packaging and electronic industries to mention a few.

The first chapter provides an overview of polymer blend technology. Chapter 2 focuses on phase separation, heterogeneous behavior, and prevention of phase separation in immiscible polymer blend systems. Crystallization and melting kinetics of phase-separated polymer blends are reported in the Chapter 3. The fourth chapter introduces and explains various theoretical studies of polymer blends. The fifth chapter summarizes different processing techniques used to prepare polymer blends. Experimental approaches for the evaluation of the phase separation process in polymer blends are reported in Chapter 6. Chapter 7 focuses on factors influencing phase separation in partially miscible polymer blends. Chapter 8 reports interfacial modification of immiscible polymer blends. Finally, Chapter 9 summarizes various industrial applications of polymer blends.

This book is ideal for polymer scientists and engineers, material scientists, researchers, and engineers, including under- and post-graduate students who are interested in this exciting field of research. It will also help industrial researchers and R&D managers who want to bring advanced phase-separated polymer materials/products into the market.

Contents

1 Introduction

1.1 Overview of the Polymer Blend Technology

Multiphasic polymer blend systems have resulted in the advent of new technologies, which are of great significance in their different applications. The combination of two different polymers has led to the tailoring of multiphasic systems possessing optimized characteristics. The blending of two polymers leads to either miscible polymer blend systems like poly(methyl methacrylate) (PMMA)/styrene-acrylonitrile copolymer (SAN) [1–3], polystyrene (PS)/poly(vinyl methyl ether) (PVME) [4, 5], and poly(methyl methacrylate) (PMMA)/poly(styrene-co-maleic anhydride) (SMA) [6] or immiscible polymer blend systems such as polycarbonate (PC)/poly(vinylidene fluoride) (PVDF) [7] and PS/PMMA [8]. This aspect will be further explained in the following chapters. Miscible polymer blend systems are a class of blends that can be fundamentally investigated to understand their equilibrium and non-equilibrium properties fully and how the demixing temperature can be altered in the presence of nanomaterials, for example, graphene oxide (GO) [9], graphene, carbon nanotubes (CNTs), and silica to mention a few. The miscibility of polymer blend systems is determined by an interaction parameter which is commonly known as the Flory–Huggins interaction parameter. The interaction parameter is mainly dependent on temperature and composition. It is important to fully understand the fundamentals behind phase separation because the resultant morphologies determine the specific application of the polymer blend system.

1.1.1 Demixing in Polymer Blend Systems

Several publications, patents, and theses about polymer blends fundamentally show their importance in different applications [10–12]. The advent of polymer blend systems, whether miscible or immiscible, resulted from the plethora of advantages they offer. Polymer blends result in multiphasic materials that have the required properties for a specific application at a low cost. It is relatively inexpensive to blend two polymers than to go into the laboratory and start synthesizing a new class of polymer with specific desired properties. Polymer blends also improve the recycling of petroleum-based polymers; it is well known that polymers have become a menace to the environment resulting in different types of pollution. This can be curbed by blending petroleum-based polymers with biodegradable polymers. Several polymers have a high glass transition temperature (T_g), resulting in them being very strenuous to process due to the elevated temperatures needed for them to flow. To ensure the effective processability of such polymers, they can be blended with polymers that have low T_g. The addition of nanomaterials into polymer blends has been a game changer because these nanomaterials induce significant effects on the properties of the polymer blends, whether immiscible or miscible. In immiscible polymer blend systems, nanomaterials can act as compatibilizers, increasing the interfacial adhesion of the two polymers; this will be further explored in **Chapter 8**. The incorporation of nanomaterials can lead to a decrease or increase in the demixing/phase separation temperature of miscible polymer blends. They can also affect the segmental motion of the polymer chains, causing either an increase or decrease in T_g and the crystallization temperature (T_c).

1.2 Different Types of Miscible Polymer Blend Systems

1.2.1 PMMA/SAN Blend Systems

PMMA/SAN is an example of a miscible polymer blend system. This polymer blend system phase separates at high temperatures and is miscible at low temperatures [13]. The miscibility of this system is a result of the copolymer repulsion effect between the acrylonitrile and phenyl units. PMMA/SAN can be prepared by solution mixing or melt mixing method. Preparation by melt mixing is the most ideal because it is the type of method which is currently being used in industry [14]. The solution mixing method is less favored because it makes use of large amounts of solvents that have detrimental effects on human health and the environment. It should also be highlighted that the number of samples that can be prepared by solution mixing is lower, and the time taken to evaporate the solvent is very long, usually requiring at least four days. Different types of nanoparticles have been reported to induce changes in polymer systems [15]. The incorporation of nanomaterials into PMMA/SAN was reported to increase the demixing temperature of the polymer blend system. Graphene oxide nanosheets, both neat and functionalized forms, were reported to increase the demixing temperature of PMMA/SAN [1, 9]. It was also reported that the addition of silica nanoparticles enhanced the demixing temperature in PMMA/SAN blend [3]. The increase in the demixing temperature was caused by the interaction of the PMMA chains with the nanoparticles; this interaction decreases the PMMA molecular weight in the bulk phase. Pang and co-workers [2] also studied the effect of nanoclay on the demixing temperature of PMMA/SAN. They observed an increase in both the binodal and the spinodal demixing temperatures after incorporating nanoclay particles in the blend systems.

1.2.2 PS/PVME Blend Systems

PS/PVME is a dynamic asymmetrical polymer blend system [16]. The T_g of PS and PVME are far apart from each other, resulting in a demixing temperature that can be observed by melt rheology. PS/PVME is another example of a low critical solution temperature (LCST) polymer blend system, as it is miscible at low temperatures but demixes at high temperatures [17]. The incorporation of different types of nanomaterials, whether functionalized or nonfunctionalized nanoparticles, was found to affect the demixing temperature of the PS/PVME blend. The addition of multi-walled CNTs (MWCNTs) was found to increase the demixing temperature of near-critical compositions. A study by Xavier et al. [5] showed that 0.25 and 0.5 wt.% of CNT increased the demixing temperature of PS/PVME as revealed by melt rheology and dielectric spectroscopy. The nanoparticle effect in the 50/50 and 60/40 blend compositions can be seen in Figure 1.1. Melt rheology is one of the widely used techniques investigating how the nanoparticles affect the demixing temperature of miscible polymer blends [6]. Other characterization techniques like optical and electron microscopes can be affected by turbidity or the presence of nanoparticles, but rheology is not affected.

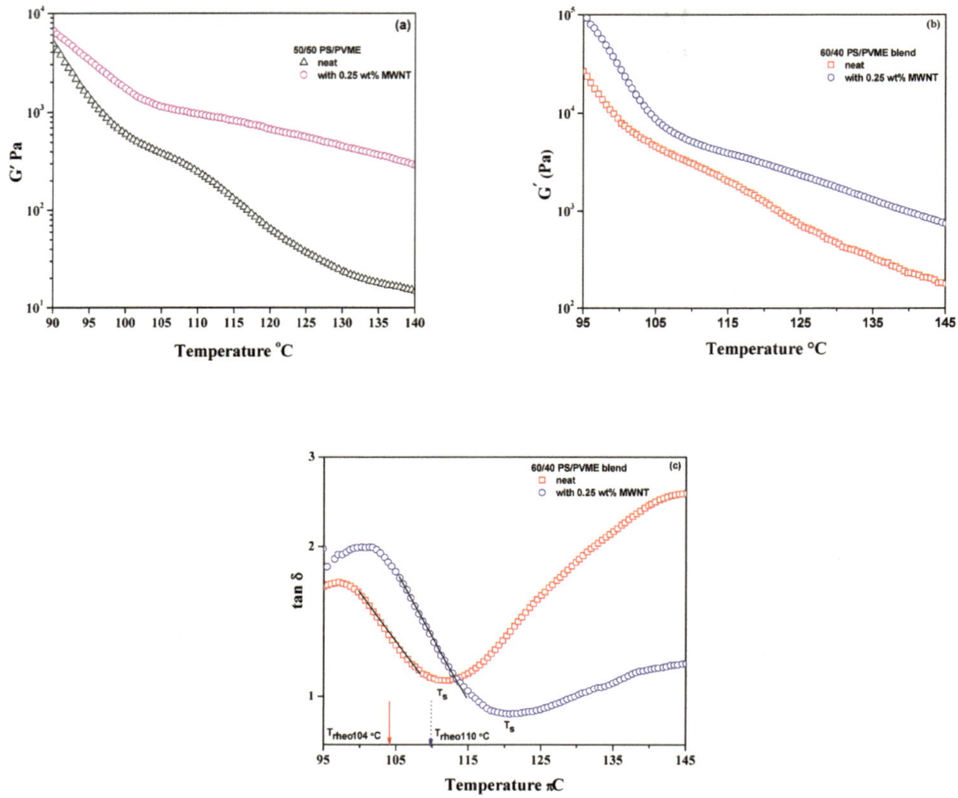

Figure 1.1 Isochronal dynamic temperature ramp performed at ω = 0.1 rad/s, 1% strain with a 0.5 °C/min heating rate for 50/50 PS/PVME blends with and without multiwalled carbon nanotubes (MWNTs) (0.25 wt.%): (a) *G*' versus temperature; and for 60/40 PS/PVME blends with and without MWNTs (0.25 wt.%), (b) *G*' versus temperature, (c) tan δ versus temperature. Reprinted with permission from [5]. Copyright 2013, American Chemical Society

Reduced GO (rGO) was also found to affect the demixing temperature of PS/PVME [18]. The incorporation of rGO increased the demixing temperature of PS/PVME. Kar and co-workers [19] went a step further in fabricating polymer-grafted nanoparticles. PS polymer chains were grafted onto the surface of silver nanoparticles and incorporated into PS/PVME polymer blend. PS polymer, which was blended with PVME, was bromo-terminated. The polymer grafted silver nanoparticles were found to increase the spinodal demixing temperature of the polymer blend system. The incorporation of spherical nanoparticles, which have a radius comparable to the radius of gyration of the polymer chains, can result in an increase in the demixing temperature. When polymer chains cover the spherical nanoparticles, there is a reduction in interaction between the two polymers (PS/PVME), increasing the demixing temperature of the polymer blend [20].

1.2.3 PVDF/PMMA Blend Systems

PVDF/PMMA is another example of a miscible polymer blend system in the melt state. The miscibility of this system is mainly a result of the dipole-dipole interaction between the CF_2 and CH_2 of PVDF and PMMA, respectively, and also due to the formation of hydrogen bonding between the CH_2 found in PVDF and the PMMA carbonyl group [21]. The PVDF/PMMA polymer system shows both LCST and upper critical solution temperature (UCST) characteristics. As explained earlier, the polymer blend is miscible at high temperatures in its melt state and demixes at low temperatures due to crystalline-induced phase separation.

■ 1.3 Immiscible Polymer Blend Systems

1.3.1 Polypropylene (PP)-Based Blend Systems

PP is one of the most used polyolefins in the world due to its low cost and availability. PP can be used to make polymer blend systems that can be used in several applications. PP can exist in three different forms atactic, syndiotactic, and isotactic. PP is widely blended with other polymers because of its outstanding thermal, mechanical, and chemical properties resulting in it being used in a number of applications, for example, automotive, medical, and packaging. An immiscible polymer blend system can be produced when PP is blended with a polar polymer, and this may usually result in poor mechanical properties. In order to improve the interfacial adhesion of PP and other polar polymers, compatibilizers can be used. The different types of compatibilizers will be extensively explored in **Chapters 7 and 8**. Polylactic acid (PLA) can be used in packaging and biomedical applications because it is a biocompatible and biodegradable polymer. Despite its ability to degrade, this polymer has several limitations, such as poor mechanical properties. These shortfalls can be prevented by blending PLA with PP to establish a PLA/PP polymer blend. The overall properties of the blend can be tailored by tuning the composition of one of the polymers, the addition of compatibilizers or nanoparticles, and the method of processing. The addition of nanoclay into the PLA/PP polymer blend system can enhance both the mechanical and thermal properties of the system resulting in the system being used in applications that require high mechanical properties and elevated temperatures [22]. Two polyolefins can also be blended with the help of a compatibilizer to improve their interfacial adhesion. Blending two polymers has resulted in tailoring multiphasic systems with high thermal conductivity. Polymers have been well known to have low thermal conductivity, and this can be improved by incorporating nanoparticles that can be localized on the interface of the immiscible blend system. Jing et al. [23] blended polyethylene (PE) and PP and incorporated hexagonal boron nitride (h-BN), which was wrapped with styrene-ethylene-propylene-styrene (SEPS) [23]. An improvement in the thermal conductivity of the blend system was observed. This can allow this system to be applied in different applications that require elevated thermal conductivity and poor electrical properties. PP can also be blended with poly(ε-caprolactone) (PCL) to improve its degradability. PP can be oxidized to improve its compatibility with PCL. In a study conducted by Arcana and co-workers [24], oxidized PP was blended with PCL via the solution casting method. The degradability of the polymer blend system was shown to improve as the PCL content was increased.

1.3.2 PE-Based Polymer Blend Systems

PE is a polymer that causes different types of pollution when dumped in the environment. In an endeavor to improve its environmental impact through fast degradation, PE has been blended with different types of biodegradable polymers. Starch, a natural polymer derived from different plants, is one of the biobased biodegradable polymers that has been recently blended with PE to reduce its environmental impact. Low-density PE (LDPE) and starch were blended, and it was observed that increasing the amount of starch in the blend system improved its degradability [25]. In another study, PE was blended with different biodegradable natural polymers, including starch, dextrin, modified wood flour, nonmodified wood flour, and lignin [26]. The main aim of the study was to investigate to what extent LDPE degrades in the presence of the above-mentioned natural polymers. It was shown an increase in the content of the natural polymers influences the degradation of LDPE.

1.3.3 PC-Based Polymer Blend Systems

PC is an engineering amorphous and transparent polymer that can be mixed with other polymers, thus taking advantage of its properties to produce polymer bend systems that can be used in different applications such as electromagnetic interference (EMI) shielding. When PC is blended with a commodity plastic such as PP, it mostly produces an immiscible polymer blend, and the interfacial adhesion can be improved by using a compatibilizer. Both PC and PP are nonconducting and for their use in EMI shielding applications different types of conducting nanoparticles must be incorporated into the system. Thomas and co-workers [27] used a PC/PP polymer blend system in EMI shielding applications. They used MWCNT to induce conductivity in the immiscible system, and as a means to compatibilize the immiscible system, and PP-grafted maleic anhydride (PP-g-MA) as a compatibilizer. A schematic of the morphological evolution in the presence of the compatibilizer and the MWCNT is shown in Figure 1.2

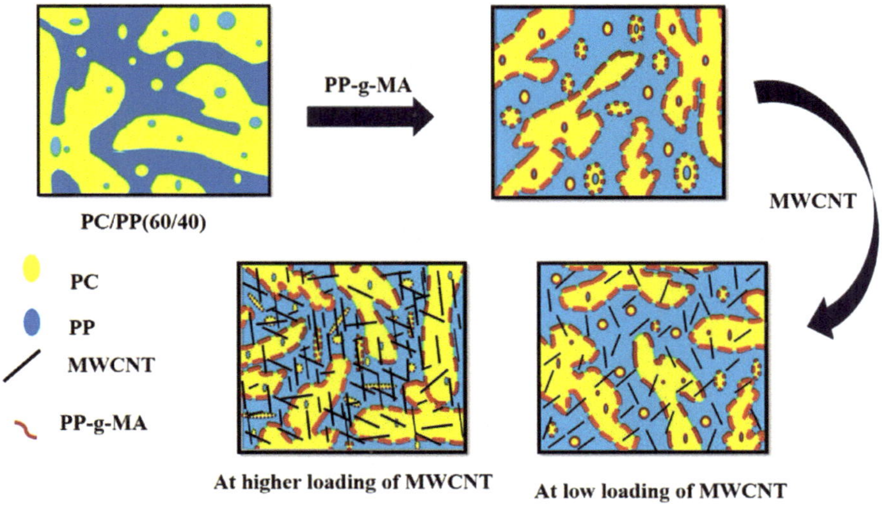

Figure 1.2 Schematic illustration of the morphological evolution of PC/PP/PP-g-MA/MWCNT nanocomposites. Reproduced with permission from [27]. Copyright 2018, American Chemical Society

The mechanisms in which immiscible polymer blends in the presence of a number of nanomaterials will be explicitly elaborated in **Chapter 8** of this book. PC can also be blended with PVDF in the presence of CNTs to make templates that can be used to attenuate electromagnetic radiation [28]. Pötschke and co-workers also blended PC with PVDF to establish a co-continuous morphology that was used in sensor applications [29]; different PCs with varying molecular weights were used. In another strategy, the PC phase was extracted, and the MWCNT was left exposed on the surface of the PVDF, as illustrated in Figure 1.3. The establishment of immiscible polymer blend systems in the presence of conducting nanomaterials is an effective method to reduce the cost of fabricating conducting composites.

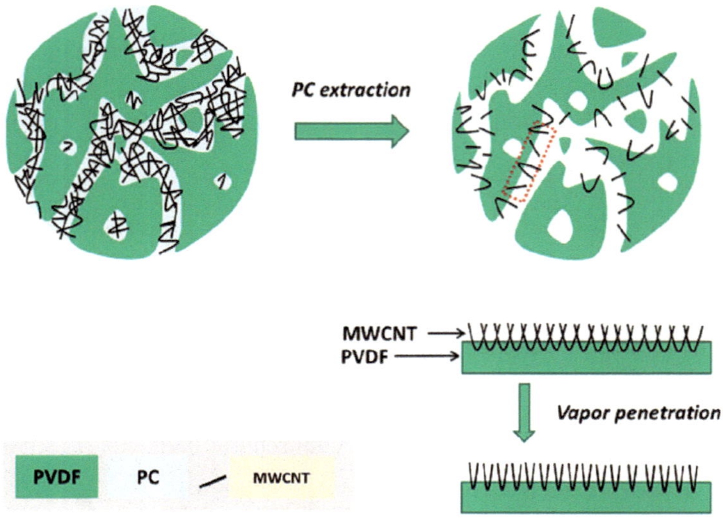

Figure 1.3 Sensing mechanism illustration of porous sensors based on co-continuous PC/PVDF/MWCNT blend composites in which PC was selectively extracted. Reproduced with permission from [29]. Copyright 2020, American Chemical Society

To prevent environmental impact, recycled PC can be blended with PE, and the resultant polymer blend system can be used in different applications. Blending these two polymers will be of great importance to protect the environment and save money. To improve the compatibility of the two polymers, compatibilizers such as oxidized PE (OPE) can be used [30]. The use of OPE improves the mechanical properties of the overall blend system enabling it to be reused in other different applications. Besides OPE, CNTs can also be used in PP-based immiscible polymer blends as compatibilizers [31].

■ 1.4 Biobased Polymer Blend Systems

An increase in plastic pollution has recently resulted in researchers focusing more on biobased polymers. Biobased polymers can either be biodegradable or nonbiodegradable. Even though most biobased polymers are environmentally friendly, their main challenge is that they are very expensive to produce. It would be of great significance to develop state-of-the-art technologies that produce relatively cheap bioplastics. In a study conducted by Misra, Mohanty, and co-workers, they mixed poly(trimethylene terephthalate) (PTT), a biobased polymer, and biobased PE (BioPE) [32] and observed an improvement in the impact strength of the new biobased blend. PTT was reported to have comparable properties to that of poly(ethylene terephthalate) and poly(butylene terephthalate) [33]. PTT has several properties, such as good dimensional stability, less water absorption, and good chemical resistance, which makes it favorable to be used in different applications. PLA is also one of the widely used polymers, especially in food packaging applications. The major shortfall of PLA is that it is very expensive, and to reduce the cost, it can be blended with another biobased polymer such as starch. Both starch and PLA can be blended, taking advantage of their individual unique properties. Starch is cheap and readily available; hence can be a useful polymer in reducing the purchase and production costs of PLA when blended. Even though starch and PLA form an immiscible polymer blend system, the interfacial adhesion can be improved by making use of compatibilizers. This improves the mechanical properties of the blend resulting in the blend being used in different applications. Different types compatibilizers of used in compatibilizing PLA/starch polymer blend can be found in Table 1.1.

Table 1.1 Different Types of Compatibilizers Used in PLA/Starch Blend Systems

Compatibilizer	Reference
Methylene diphenyl diisocyanate	[34]
Dioctyl maleate	[35]
Poly(vinyl alcohol)	[36]
Poly(hydroxyester-ether) (PHEE)	[37]
Poly(lactide)-graft-glycidyl methacrylate (PLA-g-GMA)	[38]
Maleic anhydride (MA)	[39]

Chitosan is another abundant biobased polymer that has been used in packaging applications. Its main disadvantage is that it readily absorbs moisture, and this shortfall can be repressed by mixing it with a polymer that has good moisture barrier properties. Blending chitosan with a hydrophobic polymer such as PLA can significantly improve its applications [40].

■ 1.5 Industrial Significance and Commercial Applications of Different Polymer Blend Systems

The performance of polymer blends in different applications is mainly determined by the properties of the two blended polymers, the presence of nanomaterials, and the resultant morphology of the phase-separated polymer blend system. The application of different types of polymer blend systems will be extensively investigated in **Chapter 9** of this book. In this section, a summary in table form of different kinds of polymer blends and their state-of-the-art applications are presented in Table 1.2.

Table 1.2 Recent State-of-the-Art Applications of the Polymer Blend Systems

Blend System	Application	Reference
PS/PVME	Electromagnetic shielding	[41]
PS/PVME	Antibacterial membranes	[42]
PVDF/PMMA	Suppressing electromagnetic radiation shielding	[43]
Poly(3-hexylthiophene-2,5-diyl) (P3HT) and (PMMA)	Nitrogen dioxide sensors	[44]
PVDF/PMMA	Water remediation	[45]
Polyacrylonitrile(PAN)/polyurethane (PU)	Water purification	[46]
Poly(dimethylsiloxane)/polystyrene	Optical oxygen sensors	[47]
Polypropylene (PP)/poly(ethylene terephthalate) (PET)	Electromagnetic radiation shielding	[48]
Polyvinylpyrrolidone/hydroxy-propyl-methylcellulose acetate succinate	Drug delivery	[49]
Poly(L-lactide-co-D,L-lactide)/ poly(acrylic acid)	Tissue engineering	[50]

■ 1.6 Conclusions

Polymer blends can be categorized into miscible and immiscible polymer blends depending on the specific interaction between the two polymers. The blending of two polymers has been shown to have various advantages, such as the development of materials with the desired set of properties at a relatively low cost. The effect of different types of nanomaterials on the demixing temperature, segmental dynamics, and morphology evolution of different types of polymer blends was explored in this introductory chapter. The advent of polymer blends has played a significant role in developing multiphasic structures that can be used in different state-of-the-art applications such as water remediation, drug delivery, and electromagnetic radiation shielding, to mention a few. These applications are of great importance in improving the lifestyle of humankind and in improving existing technologies. The following chapters will explicitly dwell on different types of phase separations and the resultant morphologies. Other chapters will focus more on the different types of characterization used in polymer blends, factors affecting phase separation, and different applications of phase-separated polymer blend systems.

References

[1] Muzata, T.S., Bose, S. Polymer tethered graphene oxide influences miscibility and cooperative relaxation in LCST blends. *Polymer* 188, 122127 (2020)

[2] Pang, D.Z., Zuo, M., Zhao, J.S., Zheng, Q. Influence of clay on the morphology and phase separation behavior of poly(methyl methacrylate)/poly(styrene-co-acrylonitrile) blends. *Chin. J. Polym. Sci.* 31, 1470–1483 (2013)

[3] Gao, J., Huang, C., Wang, N., Yu, W., Zhou, C. Phase separation of poly(methyl methacrylate)/poly(styrene-co-acrylonitrile) blends in the presence of silica nanoparticles. *Polymer* 53, 1772–1782 (2012)

[4] Bharati, A., Xavier, P., Kar, G.P., Madras, G., Bose, S. Nanoparticle-driven intermolecular cooperativity and miscibility in polystyrene/poly(vinyl methyl ether) blends. *J. Phys. Chem. B* 118, 2214–2225 (2014)

[5] Xavier, P., Bose, S. Multiwalled-carbon-nanotube-induced miscibility in near-critical PS/PVME blends: assessment through concentration fluctuations and segmental relaxation. *J. Phys. Chem. B* 117, 8633–8646 (2013)

[6] Li, R., Yu, W., Zhou, C. Investigation of phase separation in a partially miscible polymer blend by rheology. *J. Macromol. Sci., Part B: Phys.* 46, 1051–1062 (2007)

[7] Biswas, S., Kar, G.P., Bose, S. Attenuating microwave radiation by absorption through controlled nanoparticle localization in PC/PVDF blends. *Phys. Chem. Chem. Phys.* 17, 27698–27712 (2015)

[8] Rohini, R., Bose, S. Electromagnetic interference shielding materials derived from gelation of multiwall carbon nanotubes in polystyrene/poly(methyl methacrylate) blends. *ACS Appl. Mater. Interfaces* 6, 11302–11310 (2014)

[9] Muzata, T.S., Jagadeshvaran, P.L., Kar, G.P. Phase miscibility and dynamic heterogeneity in PMMA/SAN blends through solvent free reactive grafting of SAN on graphene oxide. *Phys. Chem. Chem. Phys.* 20, 19470–19485 (2018)

[10] Pathak, J.A., Colby, R.H., Kamath, S.Y., Kumar, S.K., Stadler, R. Rheology of miscible blends: SAN and PMMA. *Macromolecules* 31, 8988–8997 (1998)

[11] Mannella, G.A., Pavia, F.C., La Carrubba, V., Brucato, V. Phase separation of polymer blends in solution: a case study. *Eur. Polym. J.* 79, 176–186 (2016)

[12] Han, S., Zhuang, X., Shi, W., Yang, X., Li, L., Yu, J. Poly(3-hexylthiophene)/polystyrene (P3HT/PS) blends based organic field-effect transistor ammonia gas sensor. *Sens. Actuators, B* 225, 10–15 (2016)

[13] Zheng, Q., Du, M., Yang, B., Wu, G. Relationship between dynamic rheological behavior and phase separation of poly(methyl methacrylate)/poly(styrene-co-acrylonitrile) blends. *Polymer* 42, 5743–5747 (2001)

[14] Muzata, T.S., Jagadeshvaran, P.L., Bose, S. Nanoparticles influence miscibility in LCST polymer blends: from fundamental perspective to current applications. *Phys. Chem. Chem. Phys.* 22, 20167–20188 (2020)

[15] Menon, A.V., Muzata, T.S., Bose, S. Selective segregation and crystallization induced organization of carbon nanotube network in polymer nanocomposites. In *Organized Networks of Carbon Nanotubes*. Subramanian, K.R.V., Rao, A.C.L., George, R. (Eds.) (2020), Taylor & Francis, London, pp. 51–74

[16] Hajime, T. Universality of viscoelastic phase separation in dynamically asymmetric fluid mixtures. *Phys. Rev. Lett.* 76, 787-790 (1996)

[17] Shenogin, S., Kant, R., Colby, R.H., Kumar, S.K. Dynamics of miscible polymer blends: Predicting the dielectric response. *Macromolecules* 40, 5767–5775 (2007)

[18] Xavier, P. Sharma, K., Elayaraja, K., Vasu, K.S., Sood, A.K., Bose, S. Reduced graphene oxide induced phase miscibility in polystyrene-poly(vinyl methyl ether) blends. *RSC Adv.* 4, 12376–12387 (2014)

[19] Kar, G.P., Bharati, A., Xavier, P., Madras, G., Bose, S. The key role of polymer grafted nanoparticles in the phase miscibility of an LCST mixture. *Phys. Chem. Chem. Phys.* 17, 868–877 (2015)

[20] Gharachorlou, A., Goharpey, F. Rheologically determined phase behavior of LCST blends in the presence of spherical nanoparticles. *Macromolecules* 41, 3276–3283 (2008)

[21] Sharma, M., Sharma, K., Bose, S. Segmental relaxations and crystallization-induced phase separation in PVDF/PMMA blends in the presence of surface-functionalized multiwall carbon nanotubes. *J. Phys. Chem. B* 117, 8589–8602 (2013)

[22] Zawawi, E.Z.E., Hafizah, A.H.N., Romli, A.Z., Yuliana, N.Y., Bonnia, N.N. Effect of nanoclay on mechanical and morphological properties of poly(lactide) acid (PLA) and polypropylene (PP) blends. *Mater. Today: Proc.* 46, 1778–1782 (2020)

[23] Jing, X., Li, Y., Zhu, J., Maganti, S., Naik, N., Xu, B.B., Murugados, V., Huang, M., Guo, Z. Improving thermal conductivity of polyethylene/polypropylene by styrene-ethylene-propylene-styrene wrapping hexagonal boron nitride at the phase interface. *Adv. Compos. Hybrid Mater.* 5, 1090–1099 (2022)

[24] Arcana, I.M., Bundjali, B., Yudistira, I., Jariah, B., Sukria, L. Study on properties of polymer blends from polypropylene with polycaprolactone and their biodegradability. *Polym. J.* 39, 1337–1344 (2007)

[25] Abioye, A.A., Obuekwe, C.C. Investigation of the biodegradation of low-density polyethylene-starch bi-polymer blends. *Results Eng.* 5, 100090 (2020); DOI: 10.1016/j.rineng.2019.100090.

[26] Tudorachi, N., Cascaval, C.N., Rusu, M. Biodegradable polymer blends based on polyethylene and natural polymers. Degradation in soil. *J. Polym. Eng.* 20, 287–304 (2000)

[27] Poothanari, M.A., Abraham, J., Kalarikkal, N., Thomas, S. Excellent electromagnetic interference shielding and high electrical conductivity of compatibilized polycarbonate/polypropylene carbon nanotube blend nanocomposites. *Ind. Eng. Chem. Res.* 57, 4287–4297 (2018)

[28] Sushmita, K., Formanek, P., Krause, B., Pötschke, P., Bose, S. Distribution of carbon nanotubes in polycarbonate-based blends for electromagnetic interference shielding. *ACS Appl. Nano Mater.* 5, 662–677 (2022)

[29] Li, Y., Zheng, Y., Pionteck, J., Pötschke, P., Voit, B. Tuning the structure and performance of bulk and porous vapor sensors based on co-continuous carbon nanotube-filled blends of poly(vinylidene fluoride) and polycarbonates by varying melt viscosity. *ACS Appl. Mater. Interfaces* 12, 45404–45419 (2020)

[30] Ghosh, A. Recycled polyethylene/polycarbonate blends compatibilized with oxidized polyethylene/ $CaCO_3$. *J. Appl. Polym. Sci.* 139, 1–14 (2022)

[31] Poothanari, M.A., Xavier, P., Bose, S., Kalarikkal, N., Komalan, C., Thomas, S. Compatibilising action of multi-walled carbon nanotubes in polycarbonate/polypropylene (PC/PP) blends: phase morphology, viscoelastic phase separation, rheology and percolation. *J. Polym. Res.* 26, 1–15 (2019)

[32] Enriquez, E., Mohanty, A.K., Misra, M. Biobased polymer blends of poly(trimethylene terephthalate) and high density polyethylene. *Mater. Des.* 90, 984–990 (2016)

[33] Zaverl, M., Misra, M., Mohanty, A. Use of the Taguchi method for optimization of poly(butylene terephthalate) and poly(trimethylene terephthalate) blends through injection molding. *Int. Polym. Process.* 28, 454–462 (2013)

[34] Wang, H., Sun, X., Seib, P. Strengthening blends of poly(lactic acid) and starch with methylenediphenyl diisocyanate. *J. Appl. Polym. Sci.* 82, 1761–1767 (2001)

[35] Zhang, J. F., Sun, X. Mechanical and thermal properties of polylactic acid/starch blends with dioctyl maleate. *J. Appl. Polym. Sci.* 94, 1697–1704 (2004)

[36] Ke, T., Sun, X. S. Starch, poly(lactic acid), and poly(vinyl alcohol) blends. *J. Polym. Environ.* 11, 7–14 (2003)

[37] Shogren, R. L., Doane, W. M., Garlotta, D., Lawton, J. W., Willett, J. L. Biodegradation of starch/polylactic acid/poly(hydroxyester-ether) composite bars in soil. *Polym. Degrad. Stab.* 79, 405–411 (2003)

[38] Liu, J., Jiang, H., Chen, L. Grafting of glycidyl methacrylate onto poly(lactide) and properties of PLA/starch blends compatibilized by the grafted copolymer. *J. Polym. Environ.* 20, 810–816 (2012)

[39] Zhang, J. F., Sun, X. Mechanical properties of poly(lactic acid)/starch composites compatibilized by maleic anhydride. *Biomacromolecules* 5, 1446–1451 (2004)

[40] Suyatma, N. E., Copinet, A., Coma, V., Tighzert, L. Mechanical and barrier properties of biodegradable films based on chitosan and poly(lactic acid) for food packaging application. *J. Polym. Environ.* 12, 1–6 (2002)

[41] Xavier, P., Bose, S. Electromagnetic shielding materials and coatings derived from gelation of multiwall carbon nanotubes in an LCST mixture. *RSC Adv.* 4, 55341–55348 (2014)

[42] Xavier, P., Jain, S., Srinivas T, V., Chatterjee, K., Bose, S. Designer porous antibacterial membranes derived from thermally induced phase separation of PS/PVME blends decorated with an electrospun nanofiber scaffold. *RSC Adv.* 6, 10865–10872 (2016)

[43] Sharma, M., Singh, D., Menon, A., Madras, G., Bose, S. Suppressing electromagnetic radiation by trapping ferrite nanoparticles and carbon nanotubes in hierarchical nanoporous structures designed by crystallization-induced phase separation. *ChemistrySelect* 3, 1189–1201 (2018)

[44] Hou, S., Yu, J., Zhuang, X., Li, D., Liu, Y., Gao, Z., Sun, T., Wang, F., Yu, X. Phase separation of P3HT/PMMA blend film for forming semiconducting and dielectric layers in organic thin-film transistors for high-sensitivity NO_2 detection. *ACS Appl. Mater. Interfaces* 11, 44521–44527 (2019)

[45] Remanan, S., Ghosh, S., Das, T. K., Sharma, M., Bose, M., Bose, S., Das, A. K., Das, N. C. Gradient crystallinity and its influence on the poly(vinylidene fluoride)/poly(methyl methacrylate) membrane-derived by immersion precipitation method. *J. Appl. Polym. Sci.* 137, 1–20 (2020)

[46] Panda, S. R., De, S. Preparation, characterization and antifouling properties of polyacrylonitrile/polyurethane blend membranes for water purification. *RSC Adv.* 5, 23599–23612 (2015)

[47] Köse, M. E., Crutchley, R. J., DeRosa, M. C., Ananthakrishnan, N., Reynolds, J. R., Schanze, K. Morphology and oxygen sensor response of luminescent Ir-labeled poly(dimethylsiloxane)/polystyrene polymer blend films. *Langmuir* 21, 8255–8262 (2005)

[48] Sadeghi, A., Moeini, R., Yeganeh, J. K. Highly conductive PP/PET polymer blends with high electromagnetic interference shielding performances in the presence of thermally reduced graphene nanosheets prepared through melt compounding. *Polym. Compos.* 40, E1461–E1469 (2019)

[49] Tipduangta, P., Belton, P., Fabian, Wang L. Y., Tang, H., Eddleston, M., Qi, S. Electrospun polymer blend nanofibers for tunable drug delivery: The role of transformative phase separation on controlling the release rate. *Mol. Pharmaceutics* 13, 25–39 (2016)

[50] Ghaffari-Bohlouli, P., Shahrousvand, M., Zahedi, P. Performance evaluation of poly(L-lactide-co-D,L-lactide)/poly(acrylic acid) blends and their nanofibers for tissue engineering applications. *Int. J. Biol. Macromol.* 122, 1008–1016 (2019)

2 Phase Separation, Heterogeneous Behavior and Prevention of Phase Separation

2.1 Basic Concept of Polymer Blends

Blending polymers appears to be the most versatile and cost-effective way to produce polymeric materials with an extraordinary combination of properties for a wide variety of applications in electronic systems, locomotive industries, packaging industries, and household materials. Polymer blends may be described as a mixture of binary polymers or copolymers with certain compositions. Polymer blends, based on their compatibility, can be miscible (homogenous), partially miscible, and heterogeneous (immiscible), as presented in Figure 2.1 [1].

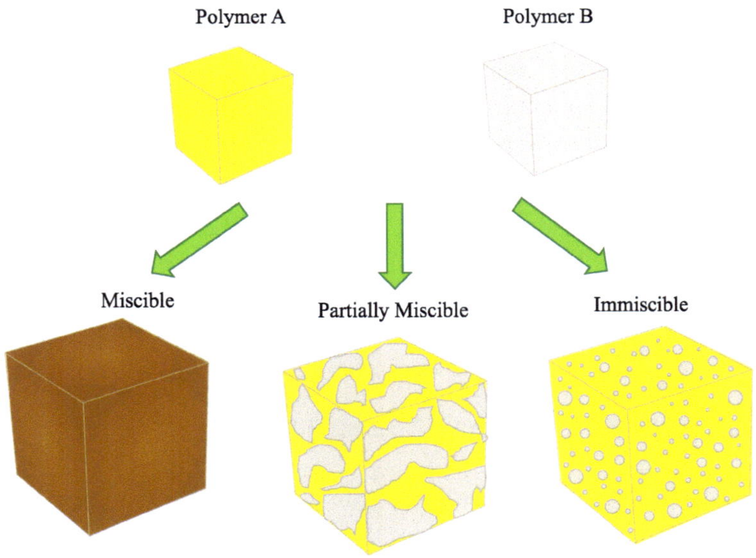

Figure 2.1 Schematic representation of categories in polymer blends: miscible, partially miscible and immiscible. Reproduced with permission from [1]. Copyright 2020, Elsevier Science Ltd

Thermodynamically, the phase behavior of binary polymer blends is regulated by Gibbs free energy of mixing. The basic thermodynamic relationship governing the mixing of polymer blends is given by Equation 2.1.

$$\Delta G_{\mathrm{m}} = \Delta H_{\mathrm{m}} - T\Delta S_{\mathrm{m}} \tag{2.1}$$

Where ΔG_{m} is the free energy of mixing, ΔH_{m} is the enthalpy of mixing (heat of mixing), and ΔS_{m} is the entropy of mixing. The blends are homogeneous (miscible at the molecular level), ΔG_{m} of Equation 2.1 must be smaller than zero or negative ($\Delta G_{\mathrm{m}} < 0$) and its second derivative with respect to volume fraction at constant pressure and temperature must be positive or greater than zero:

$$\left(\frac{\partial^2 \Delta G_{\mathrm{m}}}{\partial \phi^2}\right)_{P,T} > 0$$

Where ϕ is volume fraction [2]. From Equation 2.1, Flory and Huggins established the basic necessary requirement for evaluating the miscibility/phase behavior of polymer blends [3–6]. The entropy of mixing ΔS_{m} is always positive as it is a measure of randomness. Hence, the most decisive parameter that decides miscibility is ΔH_{m}. From this context, ΔG_{m} will be negative, providing that ΔH_{m} is negative, which is an exothermic reaction and requires specific interactions between the components of the blend. The specific interaction such as dipole-dipole, hydrogen bonding, or van der Waals interactions determines the miscibility of binary polymer blends. For example, PVDF and PMMA are miscible due to dipole-dipole interaction. In a similar concept, PVDF is also miscible with PVAc and PTMA. However, due to their long macromolecular chains, most binary polymer blends are immiscible or phase separated. Herein, we will cover the mechanism of phase separation, morphology created, and mechanism of stabilizing the morphology.

■ 2.2 Theories of Polymer Solutions

2.2.1 The Ideal Solutions of Small Molecules

A mixture of molecules having an identical size, when there is no volume change on mixing ($\Delta V_{\mathrm{m}} = 0$), the force between the molecules is equal, where the change in internal energy becomes zero ($\Delta U = 0$), and internal energy is the same $\Delta H_{\mathrm{m}} = 0$. Hence, the change in entropy of mixing depends only on configurational entropy, as there would not be any change in vibrational, rotational, and translational entropy associated with the molecules. For small molecules, configurational entropy is derived from lattice theory.

The Boltzmann equation is given by Equation 2.2:

$$S = K \ln W \tag{2.2}$$

Where K is the Boltzmann constant and W is the number of possible distinguishable generate arrangements.

The entropy of the mixture is given by Equation 2.3 [3]:

$$S_m = -K\left[N_1 \ln x_1 + N_2 \ln x_2\right] \tag{2.3}$$

Where N_1, N_2, x_1, and x_2 are the number of solvents, number of solutes, mole fraction of solvent, and mole fraction of solute, respectively.

As the enthalpy of mixing is zero, i.e., $\Delta H_m = 0$, the free energy of mixing (ΔG_m) is given by Equation 2.4:

$$\Delta G_m = \Delta H_m - T\Delta S_m \tag{2.4}$$

Replacing Equation 2.3 into Equation 2.4 gives Equation 2.5:

$$\Delta G_m = KT\left[N_1 \ln x_1 + N_2 \ln x_2\right]$$

$$= RT\left[n_1 \ln x_1 + n_2 \ln x_2\right] \tag{2.5}$$

2.2.2 Non-ideal Polymer Solutions

The Flory–Huggins theory for polymer blends is given by:

The entropy of mixing of polymer solution (Equation 2.6):

$$\Delta S_m = -K\left[N_1 \ln \phi_1 + N_2 \ln \phi_2\right]$$

$$= -R\left[n_1 \ln \phi_1 + n_2 \ln \phi_2\right] \tag{2.6}$$

Enthalpy of mixing of polymer solution (Equation 2.7):

$$\Delta H_m = -KT\left[N_1 \chi \phi_2\right]$$

$$= -RT\left[n_1 \chi \phi_2\right] \tag{2.7}$$

Gibbs free energy of mixing of polymer solution (Equation 2.8, Equation 2.9, and Equation 2.10):

$$\Delta G_m = KT\left[N_1 \ln\phi_1 + N_2 \ln\phi_2 + N_1\chi\phi_2\right]$$
$$= RT\left[n_1 \ln\phi_1 + n_2 \ln\phi_2 + n_1\chi\phi_2\right] \tag{2.8}$$

$$\phi_1 = \frac{x_1N_1}{x_1N_1 + x_2N_2} \quad \phi_2 = \frac{x_2N_2}{x_1N_1 + x_2N_2} \tag{2.9}$$

$$\overline{\Delta G_m} = \frac{\Delta G_m}{x_1N_1 + x_2N_2}$$

$$= KT\left[\left(\frac{1-\phi_2}{x_1}\right)\ln(1-\phi_2) + \frac{\phi_2}{x_2}\ln\phi_2 + (1-\phi_2)\chi\phi_2\right]$$

$$\frac{\partial\overline{\Delta G_m}}{\partial\phi_2} = KT\left[\frac{-1}{x_1} - \frac{\ln 1-\phi_2}{x_2} + \frac{1}{x_2} + \frac{\ln\phi_2}{x_2} + (1-2\phi_2)\chi\right] \tag{2.10}$$

Replacing equations Equation 2.9 and Equation 2.10 into Equation 2.8 and differentiating gives Equation 2.11:

$$\frac{\partial^2\overline{\Delta G_m}}{\partial{\phi_2}^2} = KT\left[\frac{1}{(1-\phi_2)x_1} + \frac{1}{\phi_2x_2} - 2\chi\right] \tag{2.11}$$

For spinodal curve (Equation 2.12 and Equation 2.13):

$$\frac{\partial^2\overline{\Delta G_m}}{\partial{\phi_2}^2} = 0$$

$$KT\left[\frac{1}{(1-\phi_2)x_1} + \frac{1}{\phi_2x_2} - 2\chi\right] = 0$$

$$\frac{1}{(1-\phi_2)x_1} + \frac{1}{\phi_2x_2} = 2\chi \tag{2.12}$$

$$\chi_s = \frac{1}{2}\left[\frac{1}{(1-\phi_2)x_1} + \frac{1}{\phi_2x_2}\right] \tag{2.13}$$

Flory interaction parameter is given by Equation 2.14:

$$\chi = a + \frac{b}{T} \tag{2.14}$$

The spinodal temperature T_s would be (Equation 2.15):

$$\chi_s = a + \frac{b}{T_s} \tag{2.15}$$

Hence by substituting Equation 2.13 into Equation 2.15 gives Equation 2.16:

$$T_s = \frac{b}{\frac{1}{2}\left(\frac{1}{(1-\phi_2)x_1} + \frac{1}{\phi_2 x_2}\right) - a} \tag{2.16}$$

A critical composition can be calculated by differentiating Equation 2.13 with respect to composition (Equation 2.17, Equation 2.18, and Equation 2.19):

$$\frac{\partial \chi_s}{\partial \phi_2} = 0 \tag{2.17}$$

$$(\phi_2)_c = \frac{1}{1 + \left(x_2/x_1\right)^{1/2}} \tag{2.18}$$

$$\chi_c = \frac{1}{2}\left(\frac{1}{\sqrt{x_2}} + \frac{1}{\sqrt{x_1}}\right)^2 \tag{2.19}$$

Therefore, the critical values are:

$$(\phi_2)_c = \frac{1}{1 + \left(x_2/x_1\right)^{1/2}}$$

$$\chi_c = \frac{1}{2}\left(\frac{1}{\sqrt{x_2}} + \frac{1}{\sqrt{x_1}}\right)^2$$

$$T_c = \frac{b}{\frac{\left[1 + \left(x_2/x_1\right)^{1/2}\right]^2}{2x_2} - a}$$

Therefore for $\chi < \chi_c$, the mixture is miscible in all compositions [7].

For symmetric blends $x_2 = x_1 = x$.

$$\chi_c = \frac{2}{x}$$

$$(\phi_2)_c = \frac{1}{2}$$

$$T_c = \frac{b}{\frac{2}{x} - a}$$

The phase diagram (temperature versus composition ϕ) of a binary polymer blend with an illustration of lower critical solution temperature (LCST) and upper critical solution temperature (UCST) behavior is reported in Figure 2.2.

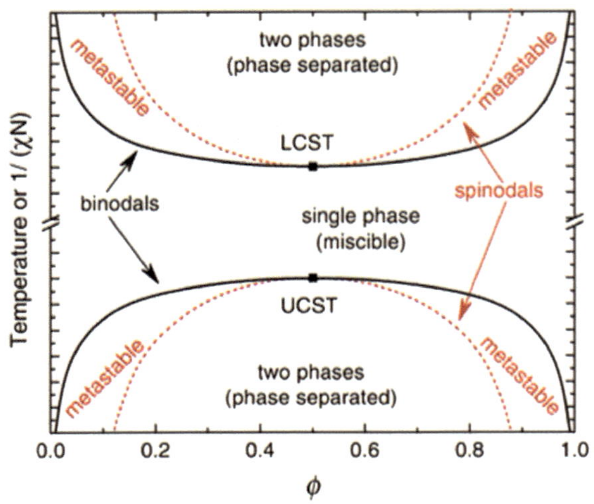

Figure 2.2 Phase diagram for binary polymer blend with an illustration of LCST and UCST behavior. **Binodal curve:** The point where the transition occurs from a single phase (miscible) into an unstable or metastable region. **Spinodal point:** The point that separates the unstable region from the metastable is called a spinodal point. **LCST:** This is the critical temperature below which the polymeric system is soluble/miscible (single phase) and above which the polymeric system shows phase separation. Phase separation occurs upon heating. The entropy of the system dominates it. **UCST:** This is the critical temperature above which polymers show a single phase (miscible) and below which blend is phase separated. Phase separation occurs upon cooling. The enthalpy of the system orchestrates them.

■ 2.3 Mechanism of Phase Separation

In binary polymer blends, there are three mechanisms by which phase separation arises, namely, nucleation and growth (NG), viscoelastic phase separation (VPS), and spinodal decomposition (SD) upon variation in composition, pressure, and temperature. The resulting morphology of phase-separated polymer blends can be droplet-matrix, matrix-fiber, lamellar, and co-continuous.

Fundamentals Theories of Nucleation and Growth

There are two types of nucleation: homogenous and heterogeneous nucleation.

Homogeneous Nucleation

Homogenous nucleation unassisted by any surface or defects. The driving force for nucleation to happen depends on Gibbs free energy (ΔG) and is given by Equation 2.20.

$$\Delta G = \Delta H - T\Delta S \tag{2.20}$$

At equilibrium:

$$\Delta G = 0$$

$$\Delta G = \Delta H - T_e \Delta S = 0$$

$$\Delta H = T_e \Delta S$$

$$\Delta G(T) = \Delta H - T\Delta S$$

$$\Delta G(T) = T_e \Delta S - T\Delta S$$

$$\Delta G(T) = \Delta S(T_e - T)$$

$$\Delta G(T) = \Delta S(\Delta T) \quad (2.21$$

Where ΔT is the supercooling temperature. For crystallization to occur, a minimum supercooling is required. Whether the crystals are growing or not depends on the following:

The total free energy would be the sum of the surface free energy ($4\pi r^2\gamma$) and volume free energy ($4/3\pi r^3\Delta G_v$).

ΔG^* and r^* are assigned to critical free energy and critical radius/size, respectively. A spherical particle can be calculated as follows (Equation 2.22):

$$\Delta G_T = -\frac{4}{3}\pi r^3 \Delta G_v + 4\pi r^2 \gamma \tag{2.22}$$

Where γ is surface tension, hence, the critical radius can be calculated by differentiating Equation 2.22 with respect to radius (Equation 2.23).

$$\frac{d\Delta G_T}{dr} = -4\pi r^2 \Delta G_v + 8\pi r\gamma = 0$$

$$r\Delta G_v = 2\gamma$$

$$r^* = \frac{2\gamma}{\Delta G_v} \tag{2.23}$$

The critical ΔG_T can be calculated by substituting Equation 2.23 into Equation 2.22.

$$\Delta G_T^* = -\frac{4}{3}\pi\left(\frac{2\gamma}{\Delta G_v}\right)^3 \Delta G_v + 4\pi\left(\frac{2\gamma}{\Delta G_v}\right)^2 \gamma$$

$$\Delta G_T^* = \frac{16\pi}{3}\frac{\gamma^3}{\Delta G_v^{\ 2}} \tag{2.24}$$

From Equation 2.23, when r is greater than the critical size r^* ($r > r^*$), the nuclei grow, and when $r < r^*$, the nuclei shrink. Preexisting particles will make nucleation easier by reducing r^* and ΔG_T^*, and hence reduce the supercooling temperature that is required for nucleation.

Heterogeneous Nucleation

Heterogeneous nucleation: nucleation is facilitated by some surface or defects (Equation 2.25 and Equation 2.26).

$$\Delta G_T = \left[-\frac{4}{3}\pi r^3 \Delta G_v + 4\pi r^2 \gamma\right] S(\theta) \tag{2.25}$$

$$S(\theta) = \frac{1}{4}(2 + \cos\theta)\left(1 - \cos^2\theta\right)^2 \tag{2.26}$$

The critical radius for heterogeneous nucleation can be found by differentiating Equation 2.25 with respect to r (Equation 2.27 and Equation 2.28).

$$\left(r^*\right)_{\text{Het}} = \frac{2\gamma}{\Delta G_v} l \tag{2.27}$$

$$\left(\Delta G_T^*\right)_{\text{Het}} = \left(\frac{16\pi}{3}\frac{\gamma^3}{\Delta G_v^{\ 2}}\right) S(\theta) \tag{2.28}$$

■ 2.4 Different Types of Morphologies

It is more economical to blend two individual polymers than to synthesize them. However, the mechanical performance of the blend is poor because of weak interfacial adhesion between two polymers. As mentioned, polymers can be miscible due to molecular chain interaction that leads to an exothermic heat of mixing or immiscible. The final properties and performance of phase-separated polymer blends are determined by crystalline morphology and structure. The morphology obtained after processing could be droplet, spherulite, lamella, fibrillar, and co-continuous. The critical parameter to tune the morphology of phase-separated polymers during processing depends on interfacial tension; the shear rate applied to the blend, the viscosity ratio between two polymers, volume fractions of polymers in the blend, and processing conditions.

Droplet Matrix and Co-continuous Morphology

As mentioned, the morphology formed during the processing of polymer blends is influenced by volume fraction, viscosity ratio, processing conditions, and interfacial tension. When the dispersed phase volume fraction is minimum or low, a dispersed droplet/matrix morphology is apparent for binary immiscible polymer blends [8]. A co-continuous morphology can be achieved upon increasing the volume fraction of the minor phase. Further increase in the extent of the minor phase leads to phase inversion, where the former dispersed phase becomes the matrix and vice versa [9]. During phase separation of binary polymeric blends, a droplet-matrix or a co-continuous morphology is generated by varying the concentration of the individual components [10]. Hence, co-continuous morphologies are developed near the phase inversion composition in a particular concentration range. From this context, Paul and Barlow [11], followed by Jordhamo and co-workers [12], proposed a governing equation based on the composition and melt viscosities of the two components for continuous, binary blends given by Equation 2.29.

$$\phi_1/\phi_2 = \eta_1/\eta_2 \tag{2.29}$$

Where, ϕ_1, ϕ_2, η_1, and η_2 are volume fraction and melt viscosity of components 1 and 2, respectively. By using Equation 2.29, the morphology of PS/PMMA, PS/*cis*-polybutadiene, and PMMA/ethylene-propylene rubber processed via melt mixing was investigated by Miles and Zurek [13]. In this direction, L. Utracki [14, 15] recently proposed Equation 2.30:

$$\lambda = \left(\frac{\phi_m - \phi_{2i}}{\phi_m - \phi_{1i}}\right)^{[\eta]\phi_m} \tag{2.30}$$

Where, ϕ_m is the matrix phase volume fraction approximated to 0.84, ϕ_{1i} and ϕ_{2i} are the continuous phase and the dispersed phase volume fractions at the phase inversion, respectively. λ is the viscosity ratio ($\lambda = \eta_1/\eta_2$). η is the intrinsic viscosity and was approximated to 1.9 [16]. The λ determines the morphology of the binary blends. When λ is larger there is coarse morphology, and if it is approximately 1 ($\lambda \sim 1$) there is fine morphology.

The impact of viscosity on phase inversion concentration for blends of PMMA/PS and PMMA/PSAN prepared by single screw extrusion was investigated by using

Equation 2.30 [17]. As a result of their interconnected/interlinked structure, co-continuous morphology has been considered for many applications where phase percolation is a significant factor which will discuss in detail in **Chapter 9** [18, 19]. In many applications, such as electrical and mechanical performance, depending on the morphology generated by phase-separated polymer blends, improving the stability and formation of co-continuous morphology is significant. It was demonstrated that the incorporation of nanofillers improves thermodynamic stability by acting as compatibilizers [20]. Several factors may contribute to the formation and stability of co-continuous morphology such as processing conditions [21], viscosity or viscosity ratio [22], blend composition [23], interfacial-tension [9], compatibilizers or graft-copolymer [24], and inorganic nanoparticles [25].

For example, PLA (polylactic acid)/PBAT (polybutylene adipate-co-terephthalate) blends are miscible due to the existence of carbonyl groups in PBAT and phase separated at PBAT content between 5 to 20 wt.% [26, 27]. Li et al. [28] analyzed the impact of blend ratio on the morphology and rheological properties of the PLA/PBAT blends. They prepared blends with different concentrations of PBAT (5, 10, and 20 wt.%) via a twin-screw extruder. Below 20 wt.% of PBAT, small droplets and uniform distribution increased viscosity and storage modulus. Moreover, when the PBAT content was between 20 to 50 wt.%, heterogeneous morphology and a decrease in viscosity and storage modulus were detected. On the other hand, when the content of PBAT was 50 wt.%, co-continuous morphology was obtained. The droplets of PLA/PBAT (20/80 w/w) blends were smaller and regular compared to PLA/PBAT (80/20 w/w) compositions. Similarly, Gu et al. [29] also reported an increase in storage moduli with an increase in PBAT content. The morphology of polymer blends can be tuned by the incorporation of nanoparticles. A change in morphology from sea-island to co-continuous morphology was observed in 50/50 w/w PPO/PA6 blends in the presence of nanoclay content greater than 5 wt.% processed via melt mixing [30]. Similar morphology transformations were observed in ABS/PA6 blends in the presence of carbon black (CB) [31]. The CB was localized in the PA6 phase and promoted the phase co-continuity to a higher composition range, resulting in a decrease in the electrical percolation threshold. The flame retardancy and thermal conductivity of immiscible PS/PVDF blends examined via localization of silicon carbide (SiC) and polystyrene (PS)-coated silicon carbide (p-SiC) nanoparticles [32]. A co-continuous morphology was observed for 30/70 vol.% PS/PVDF blend composition. When the SiC was selectively localized in PVDF (30 vol.%) phase, higher thermal conductivity was observed as compared to localizing in the PS phase. However, better flame-retardant performance was observed when p-SiC was localized in PS (30 vol.%) phase as compared to localizing p-SiC in the PVDF phase. Xiu et al. [33] examined the influence of the addition of CB on the stability of co-continuous morphology in PLA/PU (85/15 wt./wt.) blends. They found that CB with a higher capability of network formation led to stable co-continuous morphology. Similarly, the influence of annealing, incorporation of CB nanoparticles and blend composition on the microstructure and electrical property of PS/PMMA blends was investigated by Pan et al. [34]. In PS/PMMA (30/70) blend, CB was localized in the PS phase. Lower electrical percolation threshold of CB 1.26 vol.% was achieved. Interestingly, after annealing the percolation threshold of CB was decreased to 1 vol.%. At 1 vol.% CB, after annealing PS/PMMA (70/30) blends exhibited maximum conductivity.

Hence, higher electrical conductivity at a lower percolation threshold of fillers in phase-separated polymer blends is significantly important. Further reduction in percolation threshold is possible when nanoparticles are selectively localized at the interface of the co-continuous morphology of binary polymers. In this context, interfacial localization of nanoparticles in the co-continuous or interface of polymer blends is proved to be useful in stabilizing the co-continuous morphology, leading to network formation resulting in better conductivity at the minimum electrical percolation threshold. Huang et al. [35] reported that MWCNTs could systematically be localized at the interface between PLA/PCL 50/50 wt.% blends. First, MWCNTs were mixed with PLA in the melt and PCL was added. During mixing, MWCNTs migrated to the favorable PCL [35]. Mixing was stopped when the MWCNTs were localized at the interface. Consequently, the electrical percolation threshold MWCNTs was 0.97 wt.% to 0.025 wt.%. Bai et al. [36] reported a lower percolation threshold of 0.028 vol.% by localizing rGO at the interface of PLA/PS blends. This was attributed to the localization of rGO at the interface during mixing, and annealing resulted in a reduction in the coarsening of the co-continuous morphology. Shen et al. [37] studied the influence of mixing sequence in the localization of rGO (0.1–3 wt.%) in 60/40 wt.% PLA/EVA blends. The rGOs were confined in the EVA phase and at the interface when prepared from PLA/rGOs and EVA/rGOs masterbatch, respectively. They concluded that the rGOs located at the interface displayed a lower percolation threshold and resistivity as compared to rGOs located in the EVA phase. The mixing time and melt blending sequence play an important part in nanoparticle localization at the interface/co-continuous morphology of the binary components.

Bai et al. [38] reported in PLA/PS blend systems, graphene nanoplatelets (GNPs) were preferentially located at the PS phase. Interestingly, the authors were able to localize GNPs at the interface by controlling the melt blending sequence. Where GNPs were mixed first with PLA and melt blended with PS. Further, in terms of mixing time, they found that shorter melt blending time attributed to more content of GNPs trapped at the interface resulted in a reduction in coarsening. Overall, by controlling mixing time and melt blending sequence, the ability to locate GNPs at the interface increased electrical conductivity and storage modulus, as shown in Figure 2.3 [38].

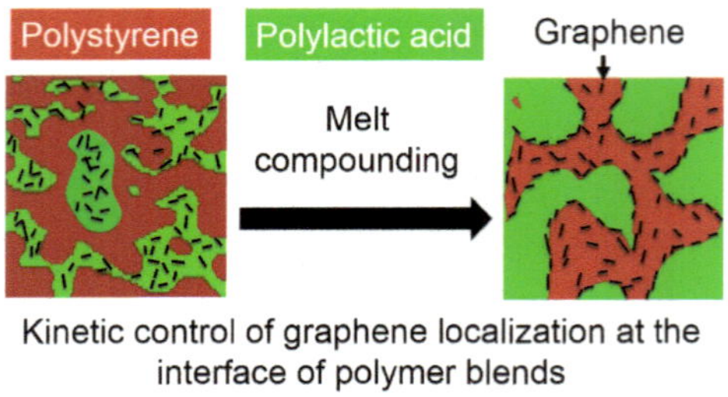

Figure 2.3 Kinetic control of graphene localization at the interface of polymer blends. Reproduced with permission from [38]. Copyright 2018, American Chemical Society

Co-continuous morphologies in binary polymers can also be stabilized by the incorporation of several types of additives with different shapes [39–41]. The critical concentration for the network formation in nanorods or nanospheres of CdSe nanoparticles in PS and poly(vinyl methyl ether) blends was investigated [42]. They found that nanorods have a lower critical concentration for network formation than nanospheres nanoparticle. In general, the impact of blend composition and viscosity ratio on the morphology of immiscible binary blends are summarized in Figure 2.4 [43].

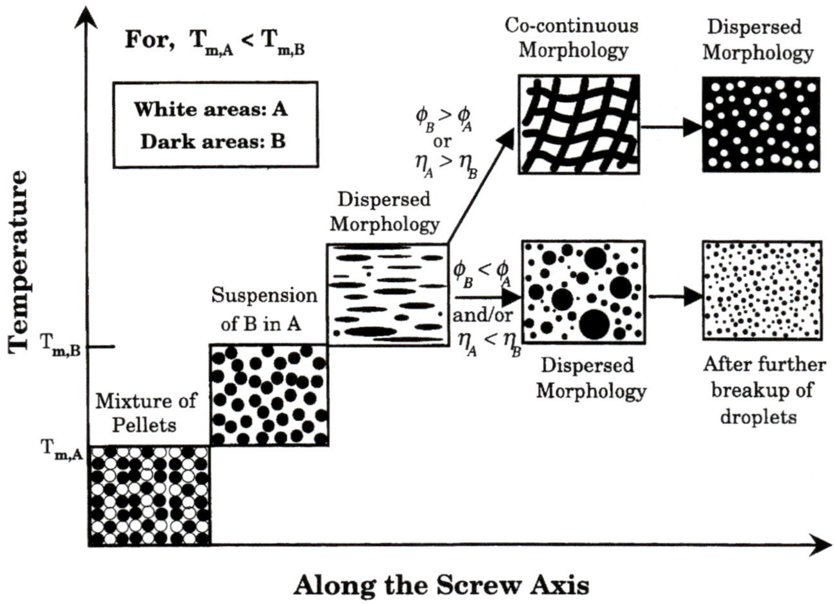

Figure 2.4 Schematic diagram describing the evolution of blend morphology of a pair of immiscible polymers, A and B, along the axis of a twin-screw extruder, where the melting point of polymer A is assumed to be lower than that of polymer B. Reproduced with permission from [43]. Copyright 1999, Elsevier Science Ltd

2.5 Influence of Compatibilizers on Morphology and the Phase Separation Behavior of Polymer Blends

The key challenges associated with phase-separated polymers are high interfacial tension, unstable/unsteady structural morphology, and weak interfacial adhesion resulting in poor mechanical performance. To overcome these challenges, it is vital to compatibilize the components of the blend. These compatibilizers are added to minimize the interfacial tension, and suppressor coalescence leads to stabilizing the blend. A wide range of approaches was adopted to compatibilize immiscible polymer blends. Herein, various strategies to compatibilize different immiscible polymer blends have been considered.

2.5.1 Influence of Reactive Compatibilization

The phenomenon of reactive compatibilization favors strategically localization and stabilization of coalescence of the droplet morphology. This can be achieved through grafting, where nanofillers interact either with one or both phases. The incorporation of a graft copolymer remarkably modified the interfacial energy between the two phases and stabilized morphology. For example, Roeder et al. [44] synthesize a compatibilizer based on PP and maleic anhydride (MA) via grafting to a compatibilized PP/polyamide-6 (PA6) blend. For the experiment, MA was grafted onto PP, and PP-g-MA was formed. The addition of 10 wt.% of PP-g-MA into 60/30 wt.% PP/PA6 showed a decrease in domain size and homogeneity of droplets. This was attributed to the specific interaction of carboxyl and amino groups of MA and PA6, respectively. Xiaomin et al. [45] assessed the effects of mixing time and amount of glycidyl methacrylate in PP-g-GMA PA on the morphology of PA1010/PP blends compatibilized with PP-g-GMA. They found that the blends with lower content of GMA showed uniform structural morphology. At the same time, phase-separated structural morphology was observed at a higher concentration of GMA. Tedesco et al. [46] fabricated different blends based on PP/nylon 6 with and without compatibilizer via melt processing at 240 °C. SEM micrographs of the different phase compositions are presented in Figure 2.5. Without compatibilizer, the nylon 6 was dispersed in PP, and poor dispersion and adhesion between phases, as shown in Figure 2.5(a). Figure 2.5(b) shows that compared to the neat PP/nylon 6 blends, a small compatibilizing effect was seen after adding PP-GMA; however, the adhesion between the phases is still poor. An effective compatibilizing effect for PP-MA was achieved, as shown in Figure 2.5(c). Hence, PP-MA is more effective in compatibilizing PP/nylon 6 as compared to PP-GMA. This was attributed to the fact that interaction between the amino group of nylon with the anhydride groups of PP-MA is more effective than epoxide groups present in PP-GMA and amino or carboxylic group present in nylon.

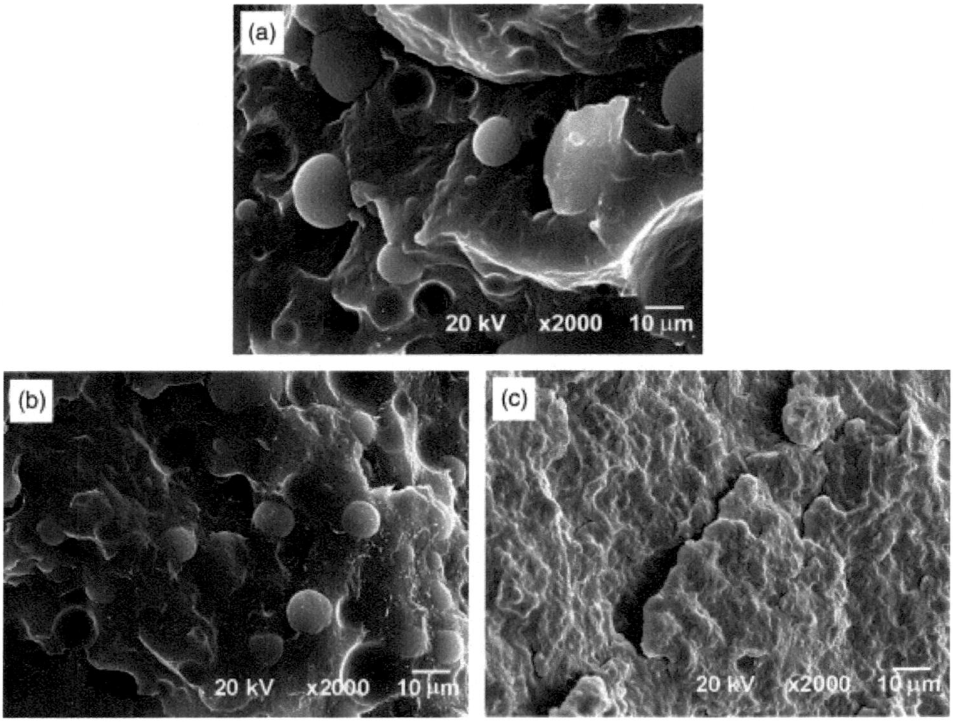

Figure 2.5 The morphology of blend (a) PP/nylon 6 (70/30); (b) PP/PP-GMA/nylon 6 (63/7/30), GMA = 0.32%; and (c) PP/PP-MA/nylon 6 (63/7/30), MA = 0.30 mol.%. Reproduced with permission from [46]. Copyright 2001, Elsevier Science Ltd

Influence of different compatibilizers (ethylene-glycidyl methacrylate (E-GMA), ethylene-acrylic acid copolymers (E-AA), ethylene-propylene copolymer grafted with maleic anhydride (EPR-g-MAH), styrene-ethylene/butylene-styrene block copolymer grafted with maleic anhydride (SEBS-g-MAH), and high-density PE (HDPE) grafted with maleic anhydride (HDPE-g-MAH)) on the morphology and chemical interactions of poly(ethylene terephthalate) (PET) and HDPE blends was investigated by Pracella et al. [47]. Among them, the E-GMA showed the highest compatibilizing effectiveness due to grafting reactions between the carboxyl/hydroxyl groups of PET and the epoxy groups of E-GMA. The compatibilizing effects of styrene-butadiene-styrene (SBS) or styrene-ethylene/butylene-styrene (SEBS) triblock copolymer to PP/PS 90/10 blends was examined [48]. With the presence of compatibilizers in PP/PS blends, low interfacial tension and reduction of the domain size of the dispersed phase. They concluded that compared to SBS, SEBS showed better compatibilizing effectiveness for PP/PS blend. Maou et al. [49] used MAH as a compatibilizer between two immiscible polymers HDPE and PVC. MAH was grafted onto the PVC and HDPE blend in the presence of dicumyl peroxide (DCP) via in situ reactive extrusion.

The grafting of MAH onto the PVC-HDPE blend was confirmed by FTIR. With the grafting of MAH, improvement in tensile strength and tensile modulus was observed, confirming an effective compatibilization effect.

2.5.2 Particle-Induced Compatibilization

In binary polymer blends, the nanoparticles are examined to be localized either in one of the phases or at the interface between the blends. Three phenomena can be observed. The nanofillers can cause suppression of the coalescence of the dispersed phases by acting as a physical barrier or can stabilize the blend structure by lowering the interfacial tension. It affected the morphology when fillers were sited in one phase by changing the viscosity ratio. For example, Gelfer et al. [50] assessed the influence of the incorporation of organoclay on the morphology of immiscible PS/PMMA blends. They found that the organoclay localized in the PMMA phase and at the interface between the two blends resulted in a reduction in microdomain sizes. The localization of nanoparticles can be tuned by the chemical functionalization of the nanoparticle. The compatibilization performance of PET/PP blends via PP-g-MA and titanium dioxide (TiO_2) was investigated [51]. TiO_2 was functionalized with polyalcohol. The treated TiO_2 reacts with PET (hydroxyl and carboxyl groups) and PP-g-MA (anhydride and hydroxyl groups). Another study examined the influence of blending sequence, and the use of PP-g-MA as a compatibilizer in PET/PP blends on the localization of TiO_2 [52]. Thermodynamically, the variation in interfacial tension between PP and TiO_2 is higher than between PET and TiO_2. Hence, the TiO_2 particles are preferred in the PET phase in the absence of PP-g-MA. Conversely, due to the presence of MA in PP-g-MA, the localization of TiO_2 was regulated by the interaction or blending sequence between TiO_2 and PP-g-MA. Xiu et al. [53] reported that the addition of TiO_2 in PLLA/PU blends and localized at the interface between the blends resulted in significant increase in impact toughness. It is well reported that the structure of graphene oxide (GO) consists of epoxy and hydroxyl groups on the basal plane of sheets and carboxyl groups at their edges [54]. The existence of these reactive groups in GO opens the way for many applications. By taking this advantage, Cao et al. [55] utilized GO sheets (GOSs) to compatibilize immiscible PA and PPO polymers. Here, it should be highlighted that the addition of small content of GOSs (0.5 wt.%) is enough to reduce the droplet size by one order of magnitude, demonstrating the compatibilizing performance of GOSs. This is due to the strong interaction between both PA and PPO with GOSs as shown in Figure 2.6. In addition, GOSs can also act as a filler to increase the mechanical and thermal properties of PA/PPO blends.

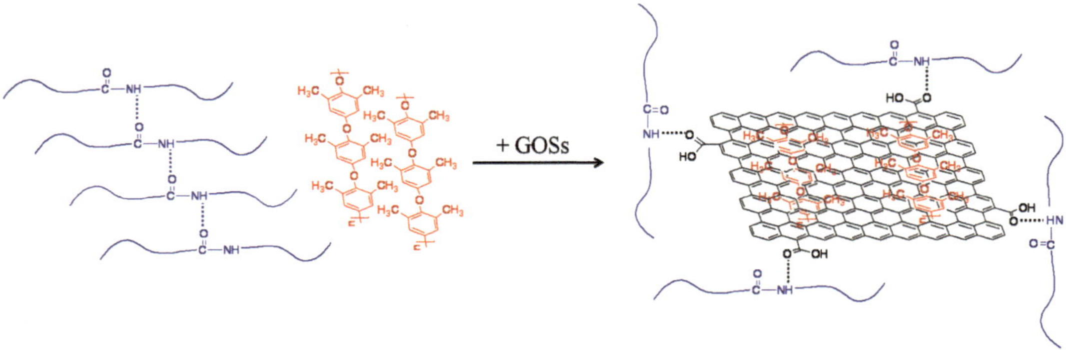

Figure 2.6 Schematic description of the compatibilizing mechanism of GOSs in the PA/PPO blends. Reproduced with permission from [55]. Copyright 2011, American Chemical Society

The incorporation of reactive nanofillers such as CNTs functionalized with different functional groups between the interface between two immiscible polymer components resulted in increased mechanical and electrical properties. For example, Chen et al. [56] proposed a mechanism to localize MWCNTs at the interface and stabilize the morphology of PS/PMMA blends. This phenomenon was attributed to the π-π interaction of PS-MWCNTs and dipole-dipole interactions of PMMA-MWCNTs components. This was probed when the carboxyl group amount in MWCNTs was approximately 0.73 wt.%. Hence the optimized carboxyl content which balances the interaction (dipole-dipole and π-π interactions interactions) is 0.73 wt.% [56]. Zhao et al. [57] fabricated functionalized CNTs, with epoxide groups on one side and long PMMA tails/chains on the opposite side, as a compatibilizer for immiscible PVDF/PLLA blends. The interactions between CF_2 present in PVDF carbonyl groups in PMMA and the carboxyl group of PLLA with epoxy groups of CNTs during melt mixing resulted in CNTs localization at the interface, as shown in Figure 2.7. Consequently, the reactive CNT compatibilized blend showed greater mechanical properties as compared to the nonreactive CNT.

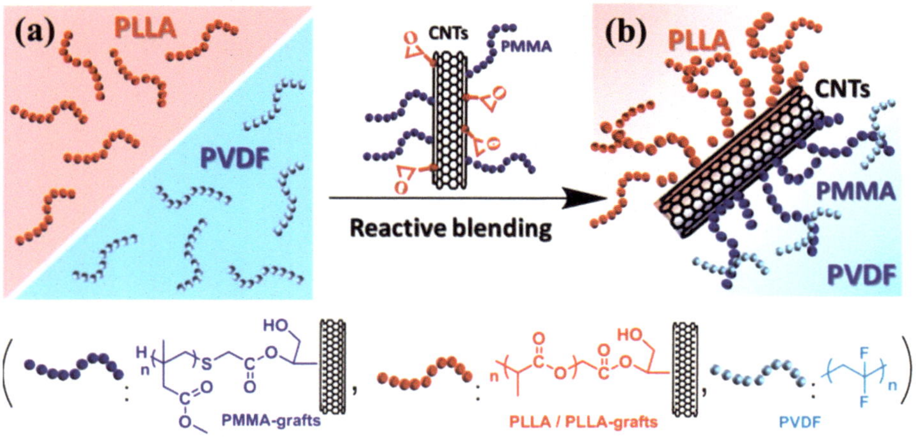

Figure 2.7 Schematic diagram of the reactive CNTs with long PMMA tails for the compatibilizer of immiscible PLLA/PVDF blends: (a) blends without CNTs with the weak interface; (b) CNTs compatibilized blends with a strengthened interface. Reproduced with permission from [57]. Copyright 2018, American Chemical Society

2.5.3 Role of Janus Hybrid Particles

Another innovative method of governing the morphology and compatibilizing effect of the immiscible polymer blend is by using Janus nanoparticles. Janus nanoparticles (JNPs) are anisotropic nanoparticles that are sectionalized into two chemically different constituent parts [58]. The localization of Janus NPs at the interphase of immiscible polymer blends showed enhanced mechanical properties of the blends [59].

Auschra and co-workers [60] synthesized different copolymers, for instance, polystyrene-b-polybutadiene-b-poly(methyl methacrylate) triblock copolymer (SBM) via anionic polymerization in THF. By taking a step further, Walther and co-workers [61] used SBM to fabricate JNPs and studied their influence on compatibilizing immiscible PS/PMMA blends, as depicted in Figure 2.8 [61]. The JNPs were localized at the interface. A substantial decrease in droplet size was observed, as proved by TEM and SEM micrographs shown in Figure 2.9. It appears to be the droplet size decreases continuously with the increased content of JNPs.

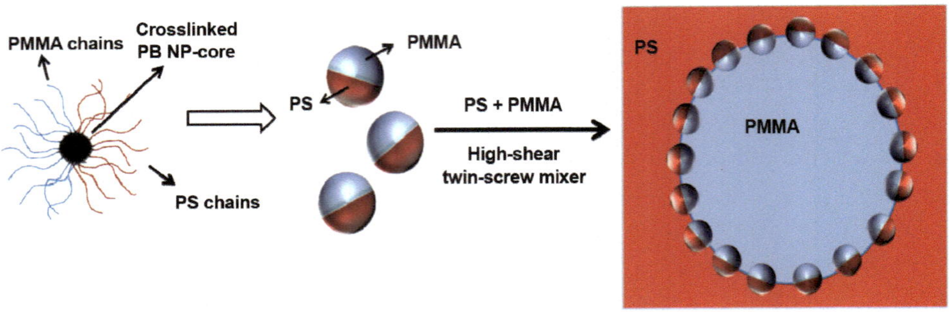

Figure 2.8 Schematic representation of Janus particles and their adsorption at the blend interface of a PS/PMMA blend. Reproduced with permission from [61]. Copyright 2008, American Chemical Society

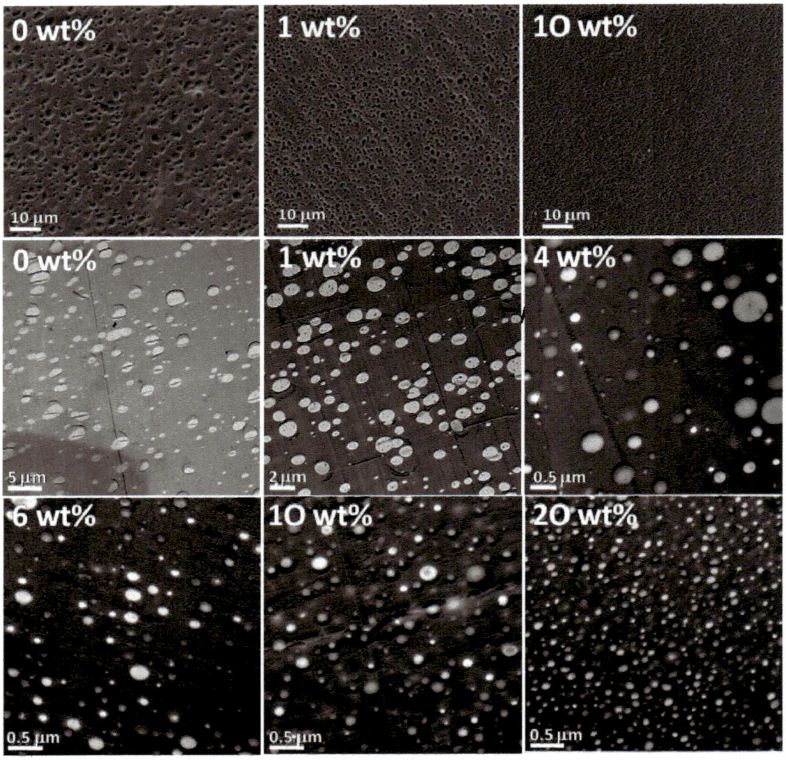

Figure 2.9 Scanning electron microscopy images obtained for blends at a PS/PMMA ratio of 6/4 (top row). Nonstained transmission electron microscopy images were obtained for blends at a PS/PMMA ratio of 8/2 (center and lower row). The amount of compatibilizer is indicated in each image. Reproduced with permission from [61]. Copyright 2008, American Chemical Society

It is well reported that the miscibility of poly(2,6-dimethyl-1,4-phenylene ether) (PPE) and poly(styrene-co-acrylonitrile) (SAN) blends count on the acrylonitrile (AN) content of the SAN phase [62]. Bahrami et al. [63] synthesized triblock terpolymer of polystyrene-b-polybutadiene-b-poly(methyl methacrylate) (SBM) Janus nanoparticles and probed the influence of the addition of different concentrations of SBM (1, 2, 5, and 10 wt.%) on the compatibilization of immiscible PPE/SAN (60/40 w/w) blends. The addition of 10 wt.% of Janus NPs resulted in a decrease in droplet size as compared with the neat one.

The compatibility of a wide range of morphologies in polymer blends can be improved by reactive blending [64]. This process favors the formation of block/graft copolymers at their interface. As a result, it lowers interface tension, reduces particle coalescence, and enhances interfacial adhesion [65–67]. Li and co-workers [68] investigated the compatibilization effect of poly(styrene-co-glycidyl methacrylate)-graft-poly(methyl methacrylate) (P((S-co-GMA)-g-MMA)) immiscible PVDF/PLLA (poly(L-lactide) blends processed via melt mixing. The epoxide groups present in P((S-co-GMA)-g-MMA) react with COOH groups of PLLA, while PVDF is miscible with PMMA due to CF_2 and carboxyl groups. Hence, the Janus nanomicelles (JNMs) are located at the interface of the PLLA/PVDF polymer blend, which acts as a compatibilizer, as shown in Figure 2.10.

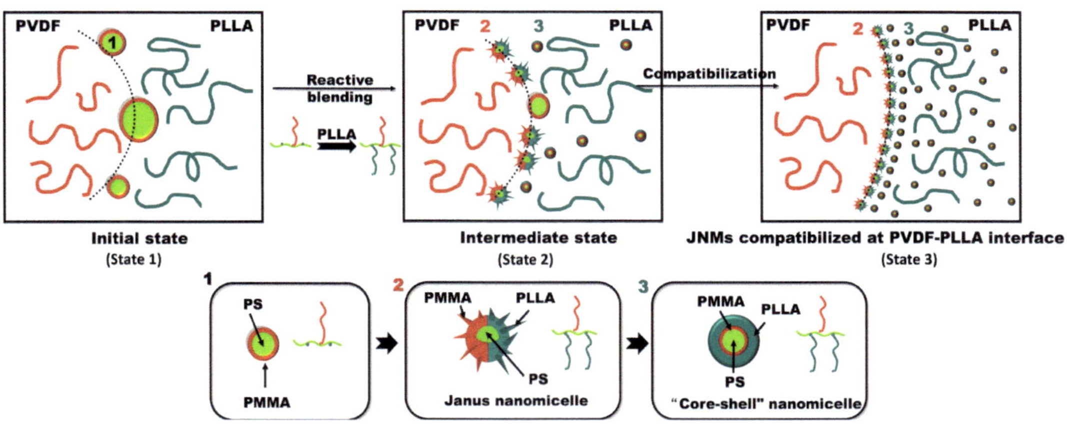

Figure 2.10 Schematic view of the in situ formation of JNMs and the morphology development of binary blends from "Droplet Stacked Domains" to "Co-continuous Encapsulated Nanomicelles" via reactive blending. Reproduced with permission from [68]. Copyright 2015, American Chemical Society

Further, the influence of molecular structure and mixing sequence of reactive graft copolymers (RGCs) poly(styrene-co-glycidyl methacrylate)-graft-poly(methyl methacrylate) on the compatibilizing performance of the immiscible PVDF/PLLA blends was demonstrated [69]. The RGCs were synthesized via copolymerization of MMA (macromer), St and GMA (monomer), as shown in Figure 2.11. First, PLLA was grafted into RGCs. This was attributed to the interaction between the carboxylic and epoxy groups in PLLA and RGCs, respectively. Second, the PMMA present on the side group RGCs reacts with PVDF due to specific interaction between CF_2 and carboxyl. Consequently, JNMs were fabricated. They were able to localize the JNMs at the interface between PLLA/PVDF blends; as a result, effective compatibilization and an increase in interfacial adhesion were observed.

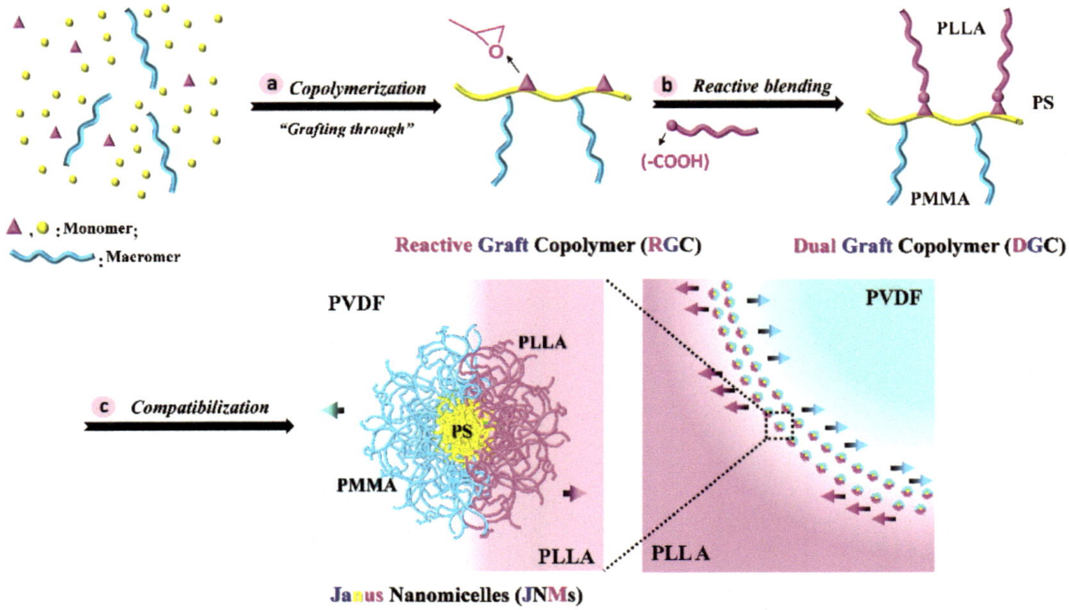

Figure 2.11 Schematic view of the in situ formation of JNMs and the proposed compatibilizing mechanism of interfacial JNMs in immiscible blends. Reproduced with permission from [69]. Copyright 2016, American Chemical Society

Another study by Yu and Wei [70] reported that the compatibilization effects of different particle shapes of JPs on immiscible blends of PLA/PC. The authors synthesized P(BA-co-AAEM)-g-SiO_2 homogenous particles and P(BA-co-AAEM)-g-SiO_2-g-PS snowman-like JPs. For synthesis, silica was modified by MPS (methacryloxypropyltrimethoxysilane), BA (*tert*-butyl acrylate), AAEM (acetoacetoxyethyl methacrylate), and St (styrene) were added during polymerization and P(BA-co-AAEM)-g-SiO_2-g-PS was achieved. They were able to localize the JPs at the interface between the two blends by controlling the surface chemistry and favored stability of the morphology. Kwon et al. [71] synthesized polymer-coated gold nanoparticles (Au NPs) with different sizes (5.9 nm and 20.7 nm) and polystyrene-b-poly(triphenylamine) (PS-b-PTPA) block copolymers and compare their compatibilizing effects in immiscible PTPA/PS blends as presented in Figure 2.12. Chen et al. [72] examined the effects of incorporation of fumed silica nanoparticles and mixing sequence in compatibilized PP/PS (20/80) blends. By controlling the mixing sequence, a reduction in PP droplet diameter from 2.7 μm to 0.7 μm was achieved.

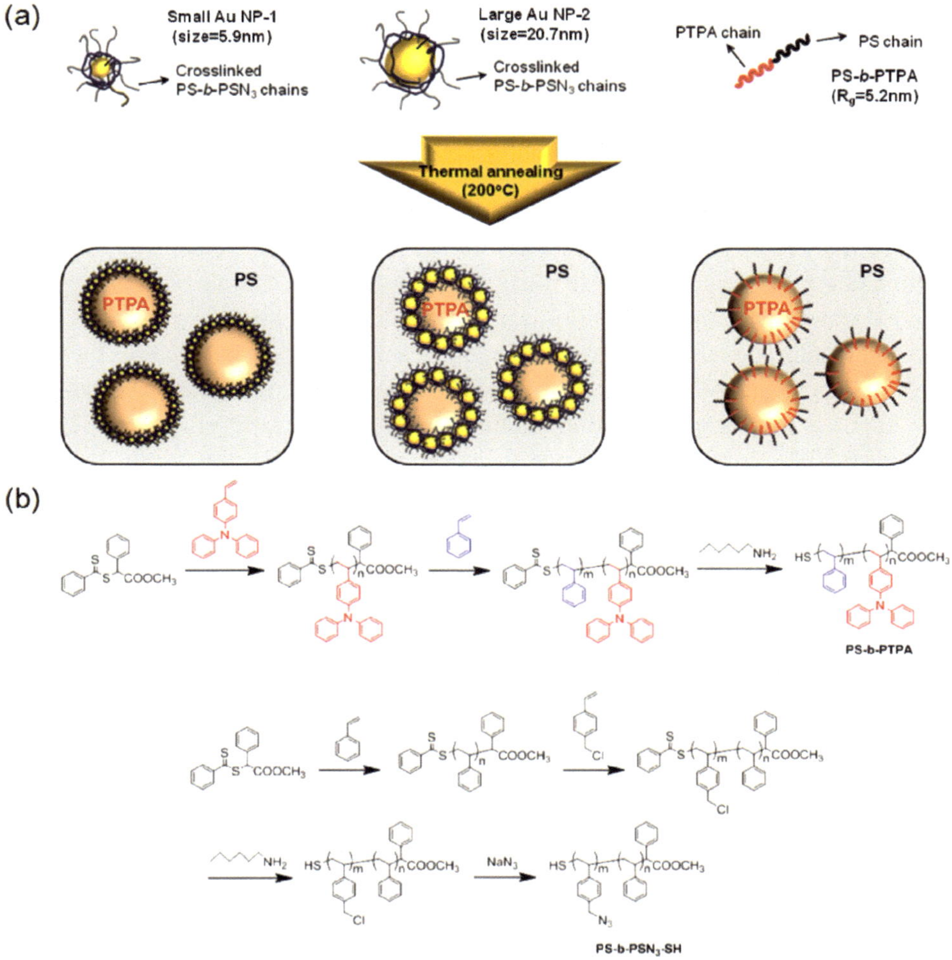

Figure 2.12 Schematic representation of small (5.9 nm) Au NP-1, large (20.7 nm) Au NP-2, and PS-b-PTPA compatibilizer in PTPA/PS blend. Reproduced with permission from [71]. Copyright 2011, American Chemical Society

2.6 Conclusions

Polymer blending is a cost-effective method to fabricate new materials with a wide variety of applications. However, the long chain and small gain in entropy of the mixing phase separate and limit their application. Phase separation occurs by nucleation and growth, viscoelastic phase separation, and spinodal decomposition upon variation in composition, pressure, and temperature. During phase separation, various morphology, such as droplet matrix morphology, matrix–fiber morphology, lamellar structures, and co-continuous morphology, can be obtained. Among them, due to their interconnected/interlinked nature, polymer blends with co-continuous morphology have been a considerable benefit for many applications where phase percolation is essential. To surmount this challenge, polymer researchers have developed compatibilizers. These compatibilizers are added to lower the interfacial tension, and suppressor coalescence leads to stabilizing the blend morphology. Tuning the location of nanoparticles in the immiscible blends offers a vast scope in the compatibilization strategy. In this context, interfacial localization of nanoparticles in co-continuous or at the interface between the binary blends is proved to be useful in stabilizing the morphologies and resulting in better conductivity at the minimum percolation threshold.

References

[1] Yong, W.F., Zhang, H. Recent advances in polymer blend membranes for gas separation and pervaporation. *Prog. Mater. Sci.* 116, 100713 (2021)

[2] Utracki, L.A. Compatibilization of polymer blends. *Can. J. Chem. Eng.* 80, 1008–1016 (2002)

[3] Flory, P.J. Thermodynamics of high polymer solutions. *J. Chem. Phys.* 10, 51–61 (1942)

[4] Flory, P.J. Thermodynamics of high polymer solutions. *J. Chem. Phys.* 9, 660–660 (1941)

[5] Huggins, M.L. Some properties of solutions of long-chain compounds. *J. Phys. Chem.* 46, 151–158 (1942)

[6] Huggins, M.L. Solutions of long chain compounds. *J. Chem. Phys.* 9, 440–440 (1941)

[7] Rubinstein, M., Colby, R.H. *Polymer Physics.* Oxford University Press, New York (2003)

[8] Favis, B.D. Factors influencing the morphology of immiscible polymer blends in melt processing. In *Polymer Blends: Formulation and Performance.* Paul, D.R., Bucknall, C.B. (Eds.) (2000), Wiley, New York, Vol. 1, 501–537

[9] Willemse, R., De Boer, A.P., Van Dam, J., Gotsis, A. Co-continuous morphologies in polymer blends: the influence of the interfacial tension. *Polymer* 40, 827–834 (1999)

[10] Vinckier, I., Laun, H.M. Manifestation of phase separation processes in oscillatory shear: droplet-matrix systems versus co-continuous morphologies. *Rheol. Acta* 38, 274–286 (1999)

[11] Paul, D., Barlow, J.W. Polymer Blends. *J. Macromol. Sci., Rev. Macromol. Chem.* 18, 109 (1980)

[12] Jordhamo, G., Manson, J., Sperling, L. Phase continuity and inversion in polymer blends and simultaneous interpenetrating networks. *Polym. Eng. Sci.* 26, 517–524 (1986)

[13] Miles, I.S., Zurek, A. Preparation, structure, and properties of two-phase co-continuous polymer blends. *Polym. Eng. Sci.* 28, 796–805 (1988)

[14] Utracki, L. A., Favis, B. Polymer alloys and blends. In *Handbook of Polymer Science and Technology*. Cheremisinoff, N. P. (Ed.), CRC Press, Boca Raton (1989), Vol. 4, 121–185

[15] Utracki, L. On the viscosity-concentration dependence of immiscible polymer blends. *J. Rheol.* 35, 1615–1637 (1991)

[16] Pötschke, P., Paul, D. Formation of co-continuous structures in melt-mixed immiscible polymer blends. *J. Macromol. Sci., Polym. Rev.* 43, 87–141 (2003)

[17] Steinmann, S., Gronski, W., Friedrich, C. Cocontinuous polymer blends: influence of viscosity and elasticity ratios of the constituent polymers on phase inversion. *Polymer* 42, 6619–6629 (2001)

[18] Gubbels, F., Blacher, S., Vanlanthem, E., Jerome, R., Deltour, R., Brouers, F., Teyssie, Ph. Design of electrical composites: determining the role of the morphology on the electrical properties of carbon black filled polymer blends. *Macromolecules* 28, 1559–1566 (1995)

[19] Willemse, R., Speijer, A., Langeraar, A., De Boer, A. P. Tensile moduli of co-continuous polymer blends. *Polymer* 40, 6645–6650 (1999)

[20] Nesterov, A., Lipatov, Y. S., Ignatova, T. Effect of an interface with solid on the component distribution in separated phases of binary polymer mixtures. *Eur. Polym. J.* 37, 281–285 (2001)

[21] Thomas, S., Groeninckx, G. Nylon 6/ethylene propylene rubber (EPM) blends: Phase morphology development during processing and comparison with literature data. *J. Appl. Polym. Sci.* 71, 1405–1429 (1999)

[22] Hedegaard, A. T., Gu, L., Macosko, C. W. Effect of extensional viscosity on cocontinuity of immiscible polymer blends. *J. Rheol.* 59, 1397–1417 (2015)

[23] Marin, N., Favis, B. D. Co-continuous morphology development in partially miscible PMMA/PC blends. *Polymer* 43, 4723–4731 (2002)

[24] Veenstra, H., van Lent, B. J., van Dam, J., de Boer, A. P. Co-continuous morphologies in polymer blends with SEBS block copolymers. *Polymer* 40, 6661–6672 (1999)

[25] Trifkovic, M., Hedegaard, A. T., Sheikhzadeh, M., Huang, S., Macosko, C. W. Stabilization of PE/PEO cocontinuous blends by interfacial nanoclays. *Macromolecules* 48, 4631–4644 (2015)

[26] Xu, Y., Hanna, M. A. Preparation and properties of biodegradable foams from starch acetate and poly(tetramethylene adipate-co-terephthalate). *Carbohydr. Polym.* 59, 521–529 (2005)

[27] Jiang, L., Wolcott, M. P., Zhang, J. Study of biodegradable polylactide/poly(butylene adipate-co-terephthalate) blends. *Biomacromolecules* 7, 199–207 (2006)

[28] Li, K., Peng, J., Turng, L. S., Huang, H. X. Dynamic rheological behavior and morphology of polylactide/poly(butylenes adipate-co-terephthalate) blends with various composition ratios. *Adv. Polym. Technol.* 30, 150–157 (2011)

[29] Gu, S.-Y., Zhang, K., Ren, J., Zhan, H. Melt rheology of polylactide/poly(butylene adipate-co-terephthalate) blends. *Carbohydr. Polym.* 74, 79–85 (2008)

[30] Li, Y., Shimizu, H. Novel morphologies of poly(phenylene oxide)(PPO)/polyamide 6 (PA6) blend nanocomposites. *Polymer* 45, 7381–7388 (2004)

[31] Wu, G., Li, B., Jiang, J. Carbon black self-networking induced co-continuity of immiscible polymer blends. *Polymer* 51, 2077–2083 (2010)

[32] Cao, J.-P., Zhao, X., Zhao, J., Zha, J-W., Hu, G-H., Dang, Z-M. Improved thermal conductivity and flame retardancy in polystyrene/poly(vinylidene fluoride) blends by controlling selective localization and surface modification of SiC nanoparticles. *ACS Appl. Mater. Interfaces* 5, 6915–6924 (2013)

[33] Xiu, H., Zhou, Y., Huang, C., Bai, H., Zhang, Q., Fu, Q. Deep insight into the key role of carbon black self-networking in the formation of co-continuous-like morphology in polylactide/poly(ether) urethane blends. *Polymer* 82, 11–21 (2016)

[34] Pan, Y., Liu, X., Hao, X., Starý, Z., Schubert, D. W. Enhancing the electrical conductivity of carbon black-filled immiscible polymer blends by tuning the morphology. *Eur. Polym. J.* 78, 106–115 (2016)

[35] Huang, J., Mao, C., Zhu, Y., Jiang, W., Yang, X. Control of carbon nanotubes at the interface of a co-continuous immiscible polymer blend to fabricate conductive composites with ultralow percolation thresholds. *Carbon* 73, 267–274 (2014)

[36] Bai, L., He, S., Fruehwirth, J. W., Stein, A., Macosko, C. W., Cheng, X. Localizing graphene at the interface of cocontinuous polymer blends: Morphology, rheology, and conductivity of cocontinuous conductive polymer composites. *J. Rheol.* 61, 575–587 (2017)

[37] Shen, Y. et al. Selective localization of reduced graphene oxides at the interface of PLA/EVA blend and its resultant electrical resistivity. *Polym. Compos.* 38, 1982–1991 (2017)

[38] Bai, L., Sharma, R., Cheng, X., Macosko, C. W. Kinetic control of graphene localization in co-continuous polymer blends via melt compounding. *Langmuir* 34, 1073–1083 (2018)

[39] Zhang, Q., Yang, H., Fu, Q. Kinetics-controlled compatibilization of immiscible polypropylene/polystyrene blends using nano-SiO_2 particles. *Polymer* 45, 1913–1922 (2004)

[40] Wu, K. H., Lu, S. Y. Preferential partition of nanowires in thin films of immiscible polymer blends. *Macromol. Rapid Commun.* 27, 424–429 (2006)

[41] Vermant, J., Cioccolo, G., Golapan Nair, K., Moldenaers, P. Coalescence suppression in model immiscible polymer blends by nano-sized colloidal particles. *Rheol. Acta* 43, 529–538 (2004)

[42] Li, L., Miesch, C., Sudeep, P. K., Balazs, A. C., Emrick, T., Russell, T. P., Hayward, R. C. Kinetically trapped co-continuous polymer morphologies through intraphase gelation of nanoparticles. *Nano Lett.* 11, 1997–2003 (2011)

[43] Lee, J. K., Han, C. D. Evolution of polymer blend morphology during compounding in a twin-screw extruder. *Polymer* 41, 1799–1815 (2000)

[44] Roeder, J., Oliveira, R., Gonçalves, M., Soldi, V., Pires, A. Polypropylene/polyamide-6 blends: influence of compatibilizing agent on interface domains. *Polym. Test.* 21, 815–821 (2002)

[45] Xiaomin, Z. et al. Morphological studies of polyamide 1010/polypropylene blends. *Polymer* 39, 15–21 (1998)

[46] Tedesco, A., Barbosa, R., Nachtigall, S., Mauler, R. Comparative study of PP-MA and PP-GMA as compatibilizing agents on polypropylene/nylon 6 blends. *Polym. Test.* 21, 11–15 (2002)

[47] Pracella, M., Rolla, L., Chionna, D., Galeski, A. Compatibilization and properties of poly(ethylene terephthalate)/polyethylene blends based on recycled materials. *Macromol. Chem. Phys.* 203, 1473–1485 (2002)

[48] Macaubas, P., Demarquette, N. Morphologies and interfacial tensions of immiscible polypropylene/polystyrene blends modified with triblock copolymers. *Polymer* 42, 2543–2554 (2001)

[49] Maou, S., Meftah, Y., Tayefi, M., Meghezzi, A., Grohens, Y. Preparation and performance of an immiscible PVC-HDPE blend compatibilized with maleic anhydride (MAH) via in-situ reactive extrusion. *J. Polym. Res.* 29, 1–10 (2022)

[50] Gelfer, M. et al. Effects of organoclays on morphology and thermal and rheological properties of polystyrene and poly(methyl methacrylate) blends. *J. Polym. Sci., Part B: Polym. Phys.* 41, 44–54 (2003)

[51] Li, W., Karger-Kocsis, J., Thomann, R. Compatibilization effect of TiO_2 nanoparticles on the phase structure of PET/PP/TiO_2 nanocomposites. *J. Polym. Sci., Part B: Polym. Phys.* 47, 1616–1624 (2009)

[52] Li, W., Karger-Kocsis, J., Schlarb, A. K. Dispersion of TiO_2 Particles in PET/PP/TiO_2 and PET/PP/PP-g-MA/TiO_2 Composites Prepared with Different Blending Procedures. *Macromol. Mater. Eng.* 294, 582–589 (2009)

[53] Xiu, H. Selective localization of titanium dioxide nanoparticles at the interface and its effect on the impact toughness of poly(L-lactide)/poly(ether) urethane blends. *eXPRESS Polym. Lett.* 7, 261–271 (2013)

[54] Dideikin, A. T., Vul', A. Y. Graphene oxide and derivatives: the place in graphene family. *Front. Phys.* 6, 149 (2019)

[55] Cao, Y., Zhang, J., Feng, J., Wu, P. Compatibilization of immiscible polymer blends using graphene oxide sheets. *ACS Nano* 5, 5920–5927 (2011)

[56] Chen, J., Cui, X., Zhu, Y., Jiang, W., Sui, K. Design of superior conductive polymer composite with precisely controlling carbon nanotubes at the interface of a co-continuous polymer blend via a balance of π-π interactions and dipole-dipole interactions. *Carbon* 114, 441–448 (2017)

[57] Zhao, X., Wang, H., Fu, Z., Li, Y. Enhanced interfacial adhesion by reactive carbon nanotubes: new route to high-performance immiscible polymer blend nanocomposites with simultaneously enhanced toughness, tensile strength, and electrical conductivity. *ACS Appl. Mater. Interfaces* 10, 8411–8416 (2018)

[58] Liang, F., Shen, K., Qu, X., Zhang, C., Wang, Q., Li, J., Liu, J., Yang, Z. Inorganic Janus nanosheets. *Angew. Chem. Int. Ed.* 50, 2379–2382 (2011)

[59] Hou, Y., Zhang, G., Tang, Z., Si, Y., Song, Z., Liang, F., Yang, Z. Janus nanosheets synchronously strengthen and toughen polymer blends. *Macromolecules* 52, 3863–3868 (2019)

[60] Auschra, C., Stadler, R. Synthesis of block copolymers with poly(methyl methacrylate): P (Bb-MMA), P (EB-b-MMA), P (SbBb-MMA) and P (Sb-EB-b-MMA). *Polym. Bull.* 30, 257–264 (1993)

[61] Walther, A., Matussek, K., Müller, A.H. Engineering nanostructured polymer blends with controlled nanoparticle location using Janus particles. *ACS Nano* 2, 1167–1178 (2008)

[62] Merfeld, G., Karim, A., Majumdar, B., Satija, S., Paul, D. Interfacial thickness in bilayers of poly(phenylene oxide) and styrenic copolymers. *J. Polym. Sci., Part B: Polym. Phys.* 36, 3115–3125 (1998)

[63] Bahrami, R., Löbling, T.I., Gröschel, A.H., Schmalz, H., Müller, A.H.E., Altstädt, V. The impact of Janus nanoparticles on the compatibilization of immiscible polymer blends under technologically relevant conditions. *ACS Nano* 8, 10048–10056 (2014)

[64] Ojijo, V., Ray, S.S., Sadiku, R. Toughening of biodegradable polylactide/poly(butylene succinate-co-adipate) blends via in situ reactive compatibilization. *ACS Appl. Mater. Interfaces* 5, 4266–4276 (2013)

[65] Jeon, H.K., Macosko, C.W., Moon, B., Hoye, T.R., Yin, Z. Coupling reactions of end-vs mid-functional polymers. *Macromolecules* 37, 2563–2571 (2004)

[66] Retsos, H., Anastasiadis, S., Pispas, S., Mays, J., Hadjichristidis, N. Interfacial tension in binary polymer blends in the presence of block copolymers. 2. Effects of additive architecture and composition. *Macromolecules* 37, 524–537 (2004)

[67] Eklind, H., Schantz, S., Maurer, F., Jannasch, P., Wesslén, B. Characterization of the interphase in PPO/PMMA blends compatibilized by P(S-g-EO). *Macromolecules* 29, 984–992 (1996)

[68] Wang, H., Dong, W., Li, Y. Compatibilization of immiscible polymer blends using in situ formed Janus nanomicelles by reactive blending. *ACS Macro Lett.* 4, 1398–1403 (2015)

[69] Wang, H., Fu, Z., Dong, W., Li, Y., Li, J. Formation of interfacial Janus nanomicelles by reactive blending and their compatibilization effects on immiscible polymer blends. *J. Phys. Chem. B* 120, 9240–9252 (2016)

[70] You, W., Yu, W. Onset reduction and stabilization of cocontinuous morphology in immiscible polymer blends by snowmanlike Janus nanoparticles. *Langmuir* 34, 11092–11100 (2018)

[71] Kwon, T., Kim, T., Ali, F.B., Kang, D.J., Yoo, M., Bang, J., Lee, W., Kim, B.J. Size-controlled polymer-coated nanoparticles as efficient compatibilizers for polymer blends. *Macromolecules* 44, 9852–9862 (2011)

[72] Chen, G., Li, P., Huang, Y., King, M., Lv, Y., Yang, Q., Li, G. Hybrid nanoparticles with different surface chemistries show higher efficiency in compatibilizing immiscible polymer blends. *Compos. Sci. Technol.* 105, 37–43 (2014)

3 Crystallization and Melting Kinetics of Phase Separated Polymer Blends

3.1 Introduction

The structural properties of polymeric materials are related to various processing techniques such as injection molding, electrospinning, and melt mixing [1, 2]. During processing, the polymeric materials are exposed to shear, which influences the crystalline morphology hence the final properties [3]. Henceforth, the determination of their thermal properties is imperative to decide the final performance. In this respect, investigation of crystallization kinetics of polymers during processing enables the design of different materials with improved properties. The supreme parameters to determine crystalline kinetics of polymer blends under isothermal or non-isothermal conditions are nucleation time, density of nuclei, and growth rate. These parameters are reflected by crystallization rate (half time and rate constant) and relative crystallinity.

3.2 Crystallization Kinetics from Melt

Crystallization has three stages, starting with nuclei, then crystalline lamellae, and finally three-dimensional structure formation. The morphologies developed after processing could be spherulite or shish-kebab. Zhang et al. [4] analyzed the influence of shear time and shear rate on the crystalline morphology of poly(phenylene sulfide) (PPS). The developed morphologies were studied by a polarized optical microscope. The PPS sample was heated to 340 °C at a heating rate of 30 °C/min and cooled to 260 °C [4]; different shear rates were applied at 260 °C and it was allowed to crystallize isothermally. The results revealed that PPS showed spherulite morphology without shear and sheared at 15 s^{-1}. When the applied shear increased to 45 s^{-1} a shish-kebab-type of morphology was observed, as shown in Figure 3.1.

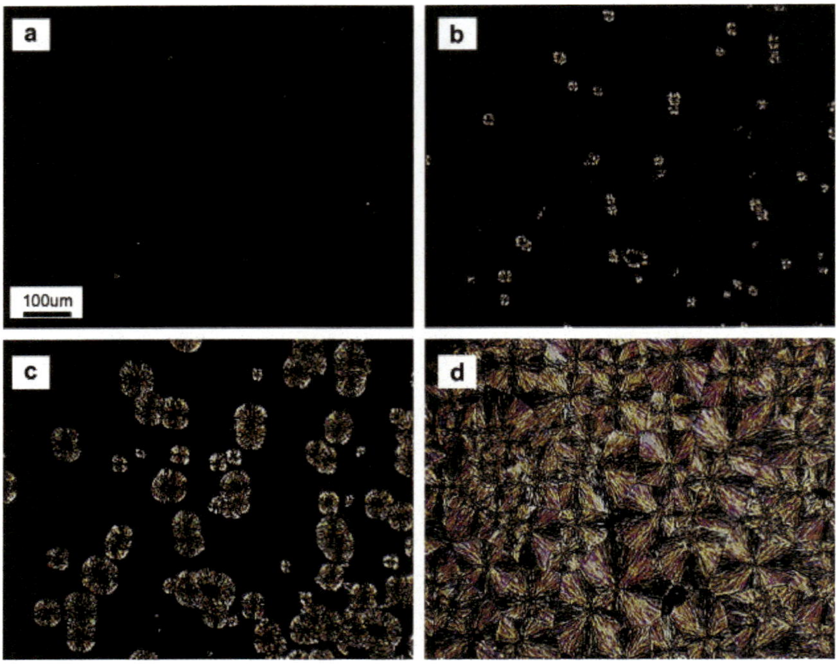

Figure 3.1 Optical micrographs of PPS during crystallization from the melt. PPS melted at 340 °C for 5 min, then quenched to 260 °C and sheared for 30 s at the shear rate of 45 s^{-1}. After shear cessation, samples were crystallized isothermally at 260 °C for (a) 18, (b) 29, (c) 55, and (d) 80 min. The black arrow indicates the flow direction. Reproduced with permission from [4]. Copyright 2008, Elsevier Science Ltd

Shear force stretched the polymer chain and led to nuclei formation resulting in shish-kebab morphology. Besides, the induction time (the time when the first crystal begins to appear) of crystallization decreases with increasing shear rate, indicating an increased rate of crystallization after shear. Here the induction time (t_0) and growth rate can be determined using rheology, differential scanning calorimeter (DSC), and polarized optical microscope (POM). Among them, rheology is superior in evaluating the flow-induced crystallization behavior of polymers crystallized isothermally or non-isothermally. Considering isothermal crystallization, the rheological time sweep experiment (storage modulus (G') with crystallization time) shows sigmoidal shapes. This signifies an increase in storage modulus reflected by three stages. Before crystallization starts (low plateau), then sharp increase during crystallization and saturation of the plateau value at the end of the crystallization process. The intersection between the vertical and horizontal line in the sigmoidal curve is ascribed as to (the induction time) signifying the energy barrier for crystallization shown in Figure 3.2.

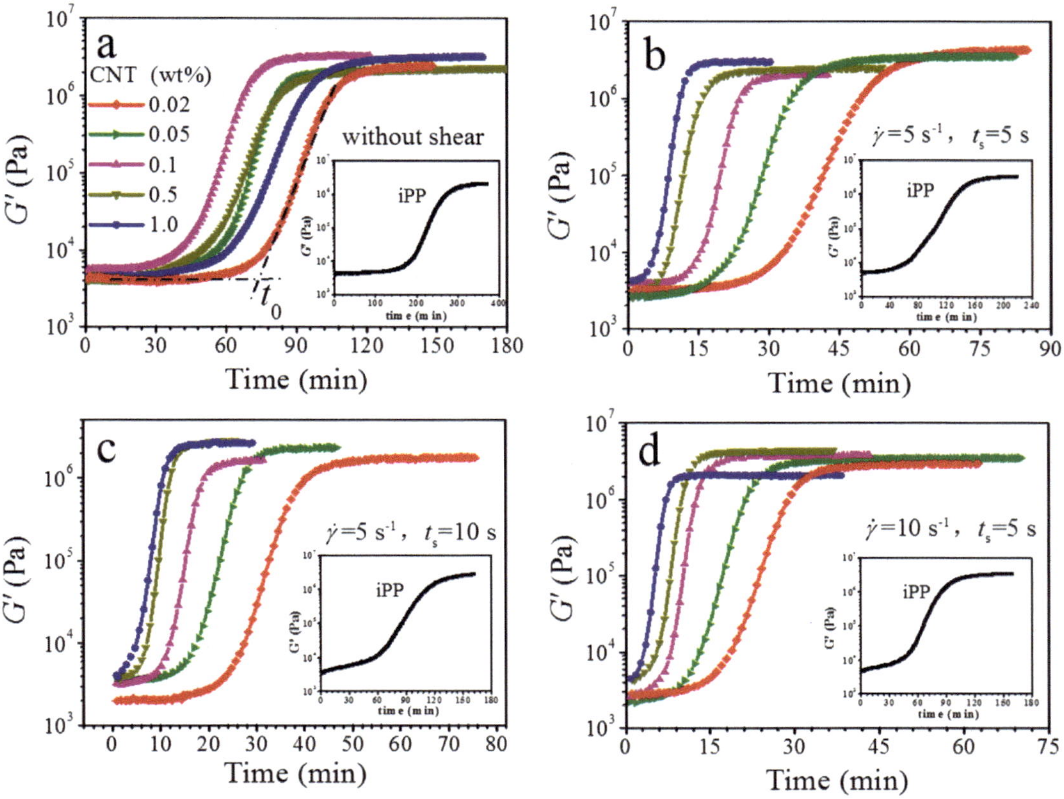

Figure 3.2 Changes of storage modulus, G', with crystallization time for neat iPP (inset) and CNT/iPP nanocomposites with different CNT contents during isothermal crystallization at T_c of 142 °C under the quiescent condition (a) and different shear conditions of (b) $\dot{\gamma}$ = 5 s^{-1}, t_s = 5 s; (c) $\dot{\gamma}$ = 5 s^{-1}, t_s = 10 s, and (d) $\dot{\gamma}$ = 10 s^{-1}, t_s = 5 s. Reproduced with permission from [5]. Copyright 2015, American Chemical Society

Wang et al. [5] used rheology to study the influence of addition of CNTs and shear flow on the crystallization kinetics of iPP/CNTs composites. A rheological time sweep experiment was conducted to determine the induction time or onset of crystallization [5]. Without shear, the minimum induction time was obtained at 0.1 wt.% CNT loading. After applying shear, the induction time was continuously decreased with increasing CNT content. Therefore, critical contents of CNTs during quiescent (without shear) conditions can be removed by applying shear, as demonstrated in Figure 3.2.

Correspondingly, Fang et al. [6] analyzed the impact of shear on isothermal crystallization kinetics of linear PLA and long chain branching PLA (LCB PLA). As shown in Figure 3.3, the sample was first heated to 200 °C, held for 5 min and cooled to 120 °C; shear was applied at 120 °C. The results proved that a significant decrease in the induction time occurred after the samples were sheared at 1 s^{-1} for both linear and branch PLA. At the same shear rate LCB PLA showed higher nuclei density and shish-kebab-type of morphology, while linear PLA showed spherulitic morphology and lower nuclei density as compared to LCB PLA. This indicates that LCB PLA has higher nucleating capacity, which is attributed to the complexity of chain relaxation.

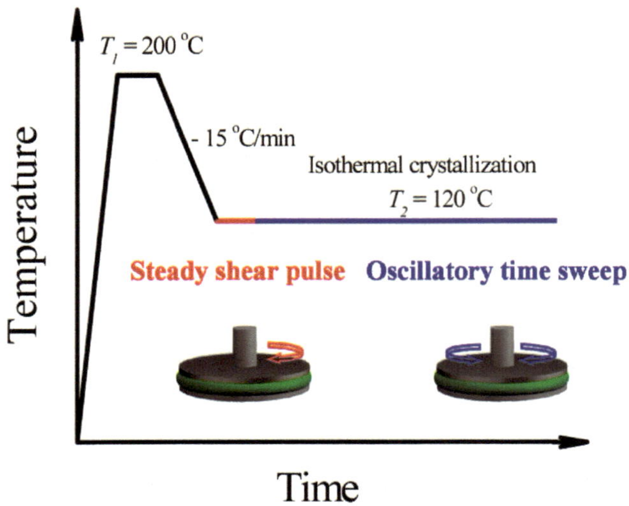

Figure 3.3
Schematic indication of thermal and shear applications for isothermal crystallization process for linear PLA and LCB PLA. Reproduced with permission from [6]. Copyright 2013, American Chemical Society

Overall, the rate of crystallization and crystalline morphology can be altered by changing shear rate, temperature, and shear time. Here, in this chapter, how nucleation occurs and factors influencing the rate of crystallization, such as blending, processing, and addition compatibilizers, will be described in detail.

■ 3.3 Theories of Nucleation and Growth

There are two types of nucleation: homogenous and heterogeneous nucleation.

3.3.1 Homogeneous Nucleation

Homogeneous nucleation is unassisted by any surface or defects. The driving force for nucleation to happen depends on Gibbs free energy (ΔG) and is given by Equation 3.1:

$$\Delta G = \Delta H - T\Delta S \tag{3.1}$$

At equilibrium (Equation 3.2):

$$\Delta G = 0$$

$$\Delta G = \Delta H - T_m \Delta S = 0$$

$$\Delta H = T_m \Delta S$$

$$\Delta G(T) = \Delta H - T\Delta S$$

$$\Delta G(T) = T_m \Delta S - T\Delta S$$

$$\Delta G(T) = \Delta S(T_m - T)$$

$$\Delta G(T) = \Delta S(\Delta T) \tag{3.2}$$

Where ΔT is the supercooling temperature [7]. In order for crystallization to occur, a minimum supercooling is required.

Whether the crystals are growing or not depends on the following: (i) The total free energy would be the sum of the surface free energy ($4\pi r^2\gamma$) and volume free energy ($4/3\pi r^3 \Delta G_v$) and (ii) the ΔG_v (volume free energy) describes the difference in energy between liquid/solid phases in bulk and is given by Equation 3.3:

$$\Delta G_v = \frac{\Delta H_f \Delta T}{T_m} \tag{3.3}$$

Where ΔH_f is the latent heat of fusion/solidification and $\Delta T = T_m - T$.

The critical radius/size (r^*) and critical free energy (ΔG^*) of the spherical cluster can be ascertained by differentiating ΔG with respect to r and setting it to zero (Equation 3.4).

$$\Delta G_T = -\frac{4}{3}\pi r^3 \Delta G_v + 4\pi r^2 \gamma \tag{3.4}$$

Where γ is surface tension.

Hence, the critical radius can be calculated by differentiating Equation 3.4 with respect to radius (Equation 3.5 and Equation 3.6).

$$\frac{d\Delta G_T}{dr} = -4\pi r^2 \Delta G_v + 8\pi r \gamma = 0$$

$$r \Delta G_v = 2\gamma$$

$$r^* = \frac{2\gamma}{\Delta G_v} \tag{3.5}$$

$$r^{*}=\frac{2\gamma T_{\mathrm{m}}}{\Delta H_{\mathrm{f}}\Delta T} \tag{3.6}$$

Equation 3.6 signifies that with greater temperature difference (ΔT) the critical radius (r^*) is smaller and, hence, crystals nucleate.

The critical ΔG_T can be calculated by substituting Equation 3.5 into Equation 3.4 giving Equation 3.7.

$$\Delta G_T^{*}=-\frac{4}{3}\pi r^{*3}\Delta G_v+4\pi r^{*2}\gamma$$

$$\left(\Delta G_T^{*}\right)_{Hom}=-\frac{4}{3}\pi\left(\frac{2\gamma}{\Delta G_v}\right)^3\Delta G_v+4\pi\left(\frac{2\gamma}{\Delta G_v}\right)^2\gamma$$

$$\left(\Delta G_T^{*}\right)_{Hom}=\frac{16\pi}{3}\frac{\gamma^3}{\Delta G_v^{\,2}} \tag{3.7}$$

From Equation 3.5, when r is greater than the critical size r^* ($r > r^*$), the nuclei grow, and when $r < r^*$, the nuclei shrink [8]. Preexisting particles will make nucleation easier by reducing r^* and $\Delta G_T{}^*$ and hence reduce the supercooling temperature required for nucleation.

3.3.2 Heterogeneous Nucleation

Heterogeneous nucleation: nucleation is facilitated by some surface or defects (Equation 3.8 and Equation 3.9).

$$\Delta G_T=\left[-\frac{4}{3}\pi r^3\Delta G_v+4\pi r^2\gamma\right]S(\theta) \tag{3.8}$$

$$S(\theta)=\frac{1}{4}(2+\cos\theta)\left(1-\cos^2\theta\right)^2 \tag{3.9}$$

The critical radius for heterogeneous nucleation can be found by differentiating Equation 3.8 with respect to r (Equation 3.10 and Equation 3.11).

$$\left(r^{*}\right)_{\mathrm{Het}}=\frac{2\gamma}{\Delta G_v} \tag{3.10}$$

$$\left(\Delta G_T^{*}\right)_{\mathrm{Het}}=\left(\frac{16\pi}{3}\frac{\gamma^3}{\Delta G_v^{\,2}}\right)S(\theta) \tag{3.11}$$

Here, $S(\theta)$ is 1 for homogenous nucleation and $0 < S(\theta) < 1$ for heterogeneous nucleation. From the above derivation, we have seen the critical radius for homogenous and heterogeneous nucleation are the same (Equation 3.5 and Equation 3.10). However, $(\Delta G_T^*)_{\mathrm{Het}}$ is lower than $(\Delta G_T^*)_{\mathrm{Hom}}$, indicating homogeneous nucleation less likely to happen as it has higher activation energy, shown in Figure 3.4 [9]. In this regard, heterogeneous nucleation is more apparently to happen as the energy barrier to prevent nucleation is lower.

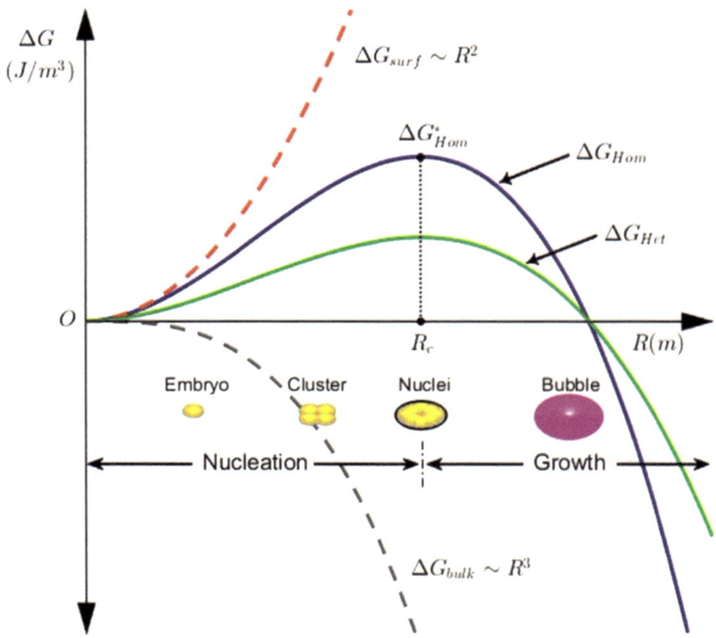

Figure 3.4 Gibbs free energy change during bubble nucleation. The surface free energy ΔG_{surf} (dashed red line), bulk-free energy ΔG_{bulk} (dashed black line), homogeneous nucleation free energy ΔG_{Hom} (blue line), and the heterogeneous nucleation free energy ΔG_{Het} (green line) are presented as a function of the radius R. The nucleation stages are also sketched for reference. The embryo and cluster states are reversible, and the nuclei state is irreversible. Beyond the nuclei phase, the bubble starts growing. Reproduced with permission from [9]. Copyright 2018, The Electrochemical Society

3.3.3 Spherulite Growth and Fold Surface Free Energy

As mentioned, when r is greater than the critical size r^* ($r > r^*$), the nuclei grow. The growth rate of polymers has been studied by the mechanism proposed by Lauritzen and Hoffman and Turnbull and Fisher [10, 11]. According to Turnbull and Fisher [10], the growth rate of spherulites is given by Equation 3.12:

$$\ln G = \ln G_0 - \frac{\Delta E^*}{kT_c} - \frac{\Delta F^*}{kT_c} \tag{3.12}$$

Where G, G_0, k, ΔE^*, ΔF^*, and T_c are the spherulite growth rate, pre-exponential factor, the Boltzmann constant, the free energy of formation of a nucleus with critical size, the free energy of activation for transporting a chain segment from the supercooled phase to the crystalline phase, and crystallization temperature, respectively.

Equation 3.12 can be simplified by considering two assumptions; that is when the T_c of a polymer approaches T_g the $-\Delta F^*/kT_c$ controls, and when T_c approaches T_m, the $-\Delta E^*/kT_c$ term controls [12]. Hence Equation 3.12 can be illustrate:

$$\ln G = \ln G_0 - \frac{\Delta E^*}{kT_c}, \quad T_c \to T_m$$

$$\ln G = \ln G_0 - \frac{\Delta E^*}{kT_c}, \quad T_c \to T_g$$

Equation 3.12 was modified by Lauritzen and Hoffman [11] for homopolymer crystallization at a large degree of supercooling and is described as follows (Equation 3.13):

$$G = G_0 \exp\left[-\frac{U^*}{R(T_c - T_\infty)}\right] \exp\left[-\frac{K_g}{T_c \Delta T f}\right] \tag{3.13}$$

Here ΔT and U^* are the degree of supercooling and the transporting activation energy, respectively. T_∞ is related to T_g and is taken when all viscous flow stops.

$$U^* = \frac{C_1 T_c}{C_2 + T_c - T_g}$$

Where C_1 and C_2 are constants.

f is a correction factor accounting for the variation in the bulk melting enthalpy per unit volume with temperature [13].

$$f = \frac{2T_c}{T_m^o + T_c}$$

K_g is a nucleation parameter, determines the region of nucleation of polymers and is correlated with lateral and chain-folded surface interfacial free energy σ and σ_e, respectively.

$$K_g\,(\text{regime I}) = \frac{4 b_0 \sigma \sigma_e T_m^o}{\Delta h_f^o k}$$

$$K_g\,(\text{regime II}) = \frac{2 b_0 \sigma \sigma_e T_m^o}{\Delta h_f^o k}$$

Where b_0 is the thickness of a monolayer and Δh_f^o is the bulk melting enthalpy per unit volume for 100% crystalline polymer. K_g is equal to K_g (I) when the rate of diffusion is higher than the rate of nucleation. K_g is equal to K_g (II) when the rate of diffusion is approximately equal to the rate of surface nucleation.

■ 3.4 Crystallization Behavior (Isothermal and Non-isothermal) of Polymer Blends

Blending polymers is being hailed as one of the smartest techniques to reduce the cost of materials. It is clear that blending may increase or decrease the crystallization kinetics of individual components depending on volume fraction, viscosity ratio, molecular weight, etc. In crystallizable polymeric materials, nucleation processes take part in a leading role. For example, Fillon et al. [14] have reported an enhanced rate of crystallization reflected by increased T_c (crystallization temperature) via the addition of different nucleating agents in isotactic polypropylene (iPP). Sun et al. [15] were able to increase the toughness strength of PBT by blending with ABS graft with glycidyl methacrylate (ABS-g-GMA). The specific reaction between the epoxy present in GMA with the COOH or OH group of PBT led to copolymer formation of ABS and PBT. As a result, uniform dispersion and enhanced tensile strength was observed at a low concentration of GMA, less than 8 wt.% in ABS.

Further, Hage et al. [16] studied the crystallization kinetics in PBT/ABS blends. They explained the addition of ABS in the blend, resulted in a reduction in the rate of nucleation of spherulite owing to high melt viscosity of acrylonitrile butadiene styrene. In contrast, the enhanced spherulitic growth rate in PBT was observed. The question appears to be how blending and the addition of nanoparticles and compatibilizers influence the crystallization kinetics and final properties during isothermal and non-isothermal crystallization.

For isothermal conditions, the rate of crystallization is determined by the Avrami equation [17–19] and is given by Equation 3.14:

$$1 - X_{(t)} = e^{-K(t)^n} \tag{3.14}$$

Where $X_{(t)}$ is the crystallinity fraction in the crystallizable material at time t. K is the crystallization rate constant and is temperature dependent; n, an integral, is the Avrami index and contains information on nucleation and growth geometry. This is under the assumption that during crystallization there is: no volume change, constant growth rate, no secondary crystallization, and constant shape and density of the growing nuclei.

By using the Avrami equation (Equation 3.14), Sencadas et al. [20] analyzed the isothermal crystallization kinetics of α-phase PVDF. The α-phase PVDF was synthesized using solution casting under DMF. They found that samples isothermally crystallized at a lower temperature (140 °C) showed higher crystallization rate, dense and small size spherulitic as compared to those isothermally crystallized at high temperature. They predicted the region of crystallization from the Lauritzen and Hoffman equation (Equation 3.13) and found α-phase PVDF crystallizes in regime III. Manna et al. [21] prepared PVDF/Ag composites via solution casting under DMF and studied the crystallization kinetics and mechanical properties. The Ag content in PVDF were 0.5, 2.5, and 11 wt.% [21]. From the Avrami equation, they found that the n value was between 1.5 and 2.0, signifying 2D nucleation and diffusion-dominated growth. Upon cooling at 5 °C/min, an increased crystallization temperature from 145.57 °C (neat PVDF) to 148.75 °C at 0.5 wt.% of Ag conforming to the nucleating effect of Ag particles for crystallization of PVDF. The specific dipole interaction between Ag nanoparticles and CF_2 led to stabilized Ag particles in the composite, increasing storage modulus at a lower concentration of Ag nanoparticles. Ye et al. [22] studied the influence of the addition of nanoparticles on crystallization kinetics in PVDF. Thin films of tetradecylphosphonic acid (TDPA)-$BaTiO_3$/PVDF with different concentrations of TDPA-$BaTiO_3$ (10, 20, and 30 wt.%) were synthesized via solution casting. An increase in β-phase formation from 66% to 93% was achieved at 20 wt.% loading. Further, owing to the nucleation effect of TDPA-$BaTiO_3$ (20 wt.%), maximum activation energy and crystallization rate were observed. On the contrary, at 30 wt.% loading, decreased crystallization rate and activation energy was shown due to the restriction of PVDF chains. Achilias et al. [23] explored the effects of the addition of SiO_2 nanoparticles with different concentrations (0.5, 2, 3.5, and 5 wt.%) in poly(propylene terephthalate) (PPT) on crystallization and nucleation kinetics via the Avrami equation. Enhanced rate of crystallization at 0.5 and 2 wt.% loading of SiO_2 nanoparticles in PPT confirm that SiO_2 is acting as nucleating agent [23]. But due to PPT chain mobility restriction at 3.5 and 5 wt.% of SiO_2 nanoparticles, the rate of crystallization decreases.

Barrau et al. [24] prepared PLA/CNTs composites via solution mixing followed by melt mixing and examined the influence of the incorporation 0.1 and 1 wt.% of CNTs in PLA on the crystallization kinetics. As depicted in Figure 3.5, the time to complete crystallization for neat PLA was 17 min and that of PLA/CNTs 8 min with 1 wt.% loading when isothermal crystallized at 110 °C. This conforms with the increased rate of crystallization of PLA with the addition of CNTs.

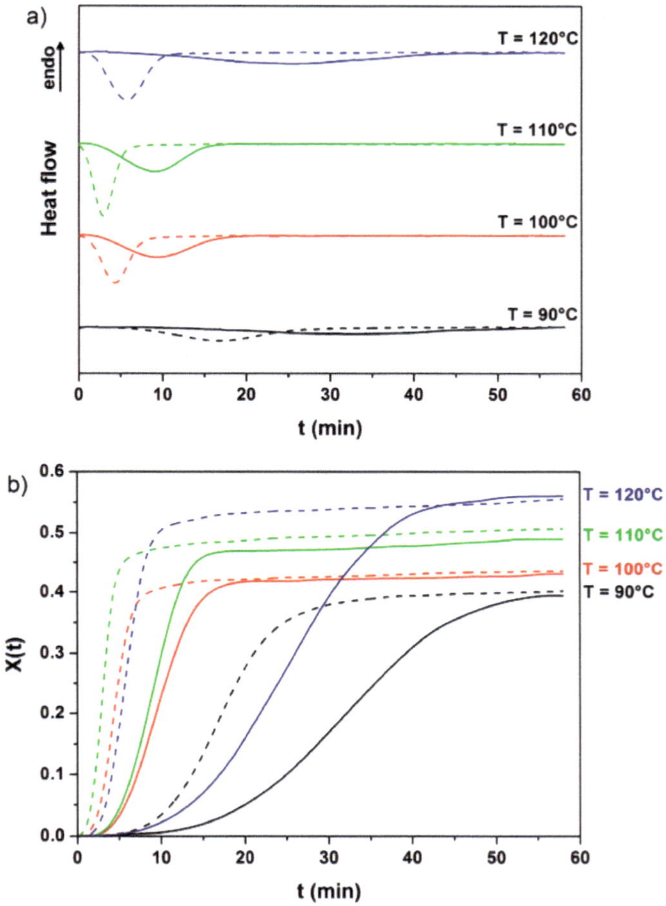

Figure 3.5 (a) Crystallization isotherms at various T_c for NT0 (solid line) and NT1 (dashed line) and (b) resulting crystallinity vs time. Reproduced with permission from [24]. Copyright 2011, American Chemical Society

Chiu et al. prepared PVDF/poly(vinyl acetate) (PVAc) blends with different concentrations (90/10, 80/20, and 70/30) via solution casting under DMF and analyzed the crystallization kinetics [25]. It was found that the Avrami exponent was around 2.5 for PVDF, indicating spherulitic morphology. In addition, the rate of crystallization and growth rate decreased with an increase in PVAc content. This confirms that the addition of PVAc restricts chain mobility of PVDF.

It is complicated to study the crystalline morphology and kinetics (growth rate and induction time) of two polymeric materials having different melting points or crystallization temperatures [26, 27]. For example, Penning and co-workers [28] studied the influence of blending two semicrystalline PVDF with PBA on the crystallization kinetics and morphology. Different blend compositions of PVDF/PBA were prepared via solution casting under DMF. From the phase diagram of PVDF/PBA, they found that the melting point of PVDF was approximately 175 °C, and that of PBA was 60 °C. During crystallization from melt, PVDF crystallized first as it had a high melting point, followed by PBA. The crystallization rate of PVDF decreases continuously with the increased content of PBA. This is attributed to the diluent effect of PBA.

Interestingly, the crystallization of PBA showed a different pattern. It reached a maximum at PVDF content of 20 wt.% and decreased as the content of PVDF further increased. They concluded the addition of PBA to PVDF altered the crystalline kinetics and polymorphism. Qiu et al. [29] studied the crystalline kinetics of crystalline-crystalline PVDF/PBSA blends. The melting point of PVDF is higher than that of PBSA. Hence, PVDF crystallizes at high temperatures while PBSA is in an amorphous state. The presence of preexisting crystals of PVDF acts as a nucleating agent for PBSA, leading to an increased rate of crystallization at lower content of PVDF. Overall, the rate of crystallization reduces with the increase in the content of PBSA in PVDF. At the same time, the nucleation rate of PBSA was increased at a low concentration of PVDF. A similar observation has been reported by Yang et al. [30], where the addition of a small amount of PVDF (20 wt.%) into PBA increases the rate of crystallization of PBA as it acts as a nucleating agent by decreasing the energy barrier for crystallization and hinders when the content of PVDF is greater than (20 wt.%) as it acts as a diluting agent. Further, the addition of small content of PVDF promoted phase transition from β- to α-phase in PBA [30]. Another study showed that, blending PVDF with PVAc, a decrease in crystallization rate and spherulite growth rate were reflected [25]. This was ascribed to the chain mobility restriction of PVDF chains due to the addition of PVAc. PVDF is miscible with PMMA at a lower concentration of PMMA. The crystallization rate of PVDF monotonically decreases with the increased amount of PMMA. This is due to the diluent effect of PMMA at higher concentrations. However, the rate of crystallization of these blends can be enhanced by applying shear force [31]. The influence of molecular weights on the crystalline morphology of binary polycaprolactone (PCL) and poly(vinyl chloride) (PVC) polymer blends was determined [32]. Different concentrations of the blends (80/20, 70/30, and 60/40 w/w) were prepared via the solution casting method under THF. As illustrated in Figure 3.6, the percentage of crystallinity of PCL monotonically decreases with an increase in the content of PVC at M_w (weight average) of PCL 1200 g/mol. This is due to the fact that that decrease in melting point upon blending led to the amount of uncrystallizable short chains. However, there was no significant influence on the percentage of crystallinity of PCL when the M_w were 15 800 g/mol and 34 800 g/mol. This is related to the chain mobility depression as increasing molecular weight. Further, in terms of chain mobility, they concluded that in 80/20 PCL/PVC blends, the mobility of chains into the interfibrillar region was promoted by decreasing the molecular weight of PCL. While in 70/30 and 60/40 blends remained unaltered by molecular weight of PCL.

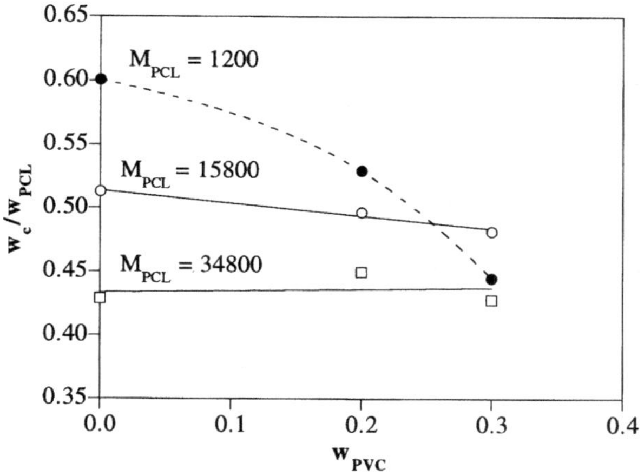

Figure 3.6 Effect of blending with PVC on the crystallizability of PCL. For the blends with low MPCL (= 1200 g/mol) the crystallizability clearly drops upon blending. For the blends with high MPCL, crystallizability is essentially unaffected. Reproduced with permission from [32]. Copyright 1998, American Chemical Society

■ 3.5 Influence of Fillers on the Morphology and Crystalline Structure of Binary Polymer Blends

The rate of crystallization and crystalline morphology in binary polymer blends depends on the selective localization of nanoparticles. In this direction, Wu et al. [33] processed PLA/PCL blends via melt mixing and investigated the localization effects of nanoclay or carbon nanotubes (CNTs) on crystalline morphology and crystallization kinetics. In order to determine the location of the nanoparticles, contact angle measurements have been conducted and the surface energy calculated. They found 1.42 $mJ{\cdot}m^{-2}$, 1.98 $mJ{\cdot}m^{-2}$, and 2.38 $mJ{\cdot}m^{-2}$ between PLA/nanoclay and PCL, PLA and PCL, and PCL/nanoclay and PLA, respectively. Due to reduced interfacial energies of PLA/nanoclay and PCL, the nanoclay was localized in the PLA phase during melt mixing (70/30 PLA/PCL/clay 3 wt.%). However, in CNTs, the viscosity ratio was the deceiving factor controlling the location of CNTs, not the thermodynamic factor hence CNTs located in the PCL phase. Interestingly, by decreasing the viscosity ratio, they were able to localize CNTs in the PLA phase. Besides, when the CNTs were localized in the PCL phase, they acted as a nucleating agent, while in PLA, increases in the rate of crystallization were noticeable. The influence of MWCNTs and maleated polypropylene (compatibilizer) (MPP) in PA6/PP on crystalline morphology and thermal and mechanical properties were examined by Zhang and co-workers [34]. 70/30 PA6/PP with different content of MWCNTs (multiwalled CNTs) (0.4, 1, and 2 parts per hundred of resin (phr)) and 5 phr MPP were prepared via melt mixing [34]. As calculated from contact angle measurements the interfacial tensions between MWNTs-A6, MWNTs-PP, and PA6-PP were 5.79 $mN{\cdot}m^{-1}$, 25.92 $mN{\cdot}m^{-1}$, and 13.5 $mN{\cdot}m^{-1}$, respectively. Thus, the MWCNTs were located at the PA6 phase due to minimum interfacial tensions. The addition of MWCNTs and MPP resulted in increased viscosity and thermal stability of the composites. Further, the MWCNTs promoted the formation of α-crystalline in PA6. However, the percentage of crystallinity of PP and PA6 decreases with the addition of MPP, owing to a reduction in chain mobility. An increase in the percentage of crystallinity from 27.7% to 42.1% was observed by the incorporation of synthetic talc particles (HT) in PP/PA6 blends [35]. The authors compared the influence of incorporation of natural (surface area 14 $m^2{\cdot}g^{-1}$) and synthetic (surface area 131 $m^2{\cdot}g^{-1}$) talc particles on the crystalline morphology, compatibilizing effectiveness, and thermal and mechanical properties in PP/PA6 blends prepared via melt processing. The samples composition was 80/20, 80/16/4, and 80/16/4 wt.% of PP/PA6, PP/PA6/natural talc particles, and PP/PA6/synthetic talc particles, respectively. Both the natural and synthetic talc particles were in the PA6 phase. They found that as compared to the natural talc particles, the synthetic talc particles showed the highest reduction in droplet domain size. They concluded that synthetic talc particles have higher compatibilization effectiveness, resulting in higher mechanical properties and percentage of crystallinity as compared to natural ones. Zou et al. [36] reported selective localization of acid functionalized MWCNTs (at low loading less than 0.5 phr) in PA66 phase in 60/40 w/w PPS/PA66 blends, co-continues morphology and uniform dispersion can be achieved as shown in Figure 3.7.

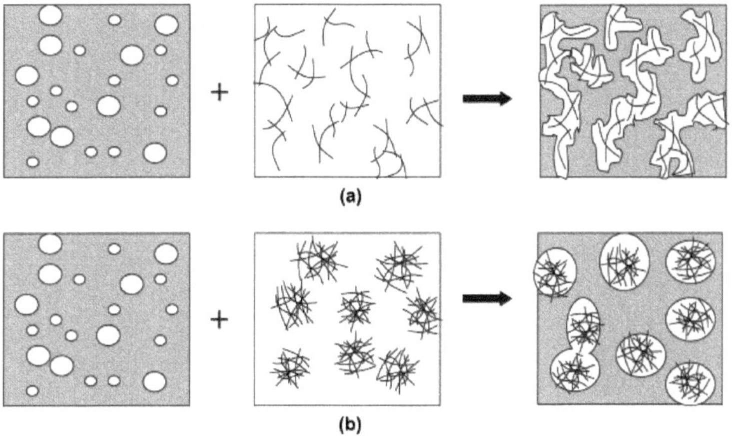

Figure 3.7 Schematic representation of the PPS/PA66 (60/40 w/w)/MWCNTs nanocomposites morphology changing: (a) low load MWCNTs (<0.5 phr) disperse evenly and form a percolating network; (b) high load MWCNTs (> 0.5 phr) aggregate-like clews. Reproduced with permission from [36]. Copyright 2006, Elsevier Science Ltd

3.6 Impact of Compatibilizers on the Crystallization Morphology and Crystallization Rate of Polymer Blends

As explained in **Chapter 2**, the blending of binary polymers is the most promising way to fabricate new material with improved properties as compared to the individual components. These properties stand on processing techniques, crystallization kinetics, and type of compatibilizers. It is well reported that the addition of compatibilizers in binary components reduces interfacial energy and improves interfacial adhesion, resulting in enhanced mechanical properties [37]. However, its plasticizing effect, which reduces the crystallization rate, limits its wide range of applications. The effects of the addition of a third component (compatibilizers) in binary polymer blends on the crystalline kinetics will focus on this section. Patra et al. [38] studied the isothermal and non-isothermal crystallization kinetics of PP/PEI blends with and without compatibilizer. The sample compositions were neat PP, PP/PP-g-MA (3 wt.%), PP/PP-g-MA (1 wt.%)/PEI (1 wt.%), and PP/PP-g-MA (3 wt.%)/PEI (3 wt.%). Among all, PP/PP-g-MA (1 wt.%)/PEI (1 wt.%) demonstrated highest degree of crystallinity and shorter time to complete crystallization. Using Avrami's model, they also found that lower crystallization half time in PP/PP-g-MA (1 wt.%)/PEI (1 wt.%) as compared to neat PP and the activation energy was 119 kJ·mol^{-1} for PP/PP-g-MA (1 wt.%)/PEI (1 wt.%) and 188 kJ·mol^{-1} for PP indicating PEI facilitating crystallization of PP in the presence of compatibilizer. The crosslink formed PEI and PP via MA resulted in an increase in T_c (crystallization temperature) [39]. Zhang et al. [40] used lactide-caprolactone copolymer (LACL) as a compatibilizer for PLA/PCL blends. The effects of blending PLA with PCL and the addition of LACL as a compatibilizer on the crystalline kinetics and morphology were analyzed. PLA/PCL/LACL samples were prepared via solution casting. A decrease in spherulite size with decrease in isothermal crystallization temperature was observed due to increased nucleation density, as shown in Figure 3.8. The density of spherulites for PLA in the presence of PCL and LACL was higher than for neat PLA. The effect is more pronounced in the presence of LACL signifying that LACL had better compatibilization and nucleation effect in PLA.

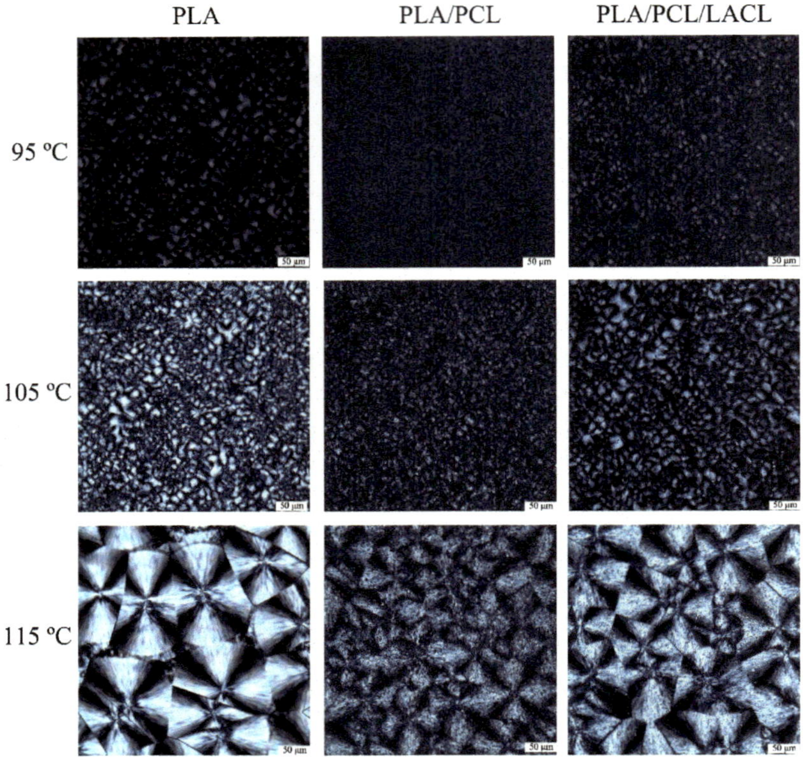

Figure 3.8 Polarized optical micrographs of PLA, PLA/PCL, and PLA/PCL/LACL spherulites isothermally melt crystallized at 95, 105, and 115 °C. Reproduced from [40]. This is an open-source article and was published under a Creative Commons Attribution 3.0 International License

Zhang et al. [41] introduced nano-sized SiO_2 to compatibilized immiscible PP/PS (70/30) blends and studied the crystallization kinetics. The domain size of PS decreases with increasing content of SiO_2 confirming the compatibilizing effect. However, the rate of crystallization of PP decreases with increasing the amount of SiO_2 due to an increase in the viscosity of the blends, as depicted in Figure 3.9 [41].

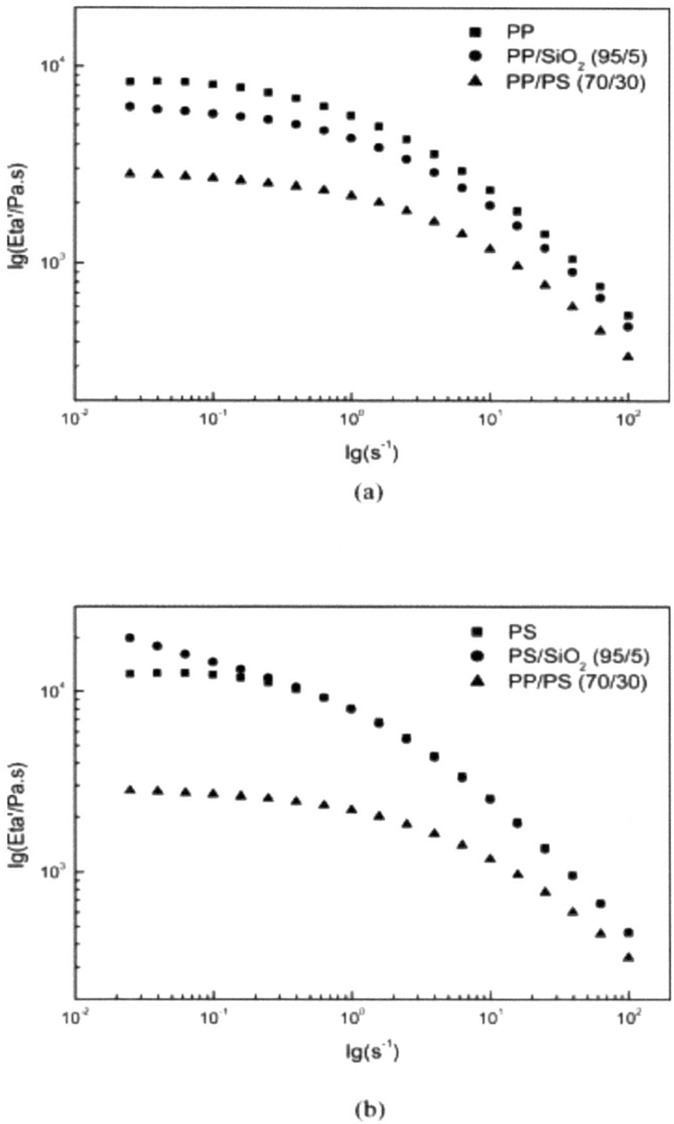

Figure 3.9 (a) Storage viscosity of PP, PP/PS(70/30), and PP/nano-SiO_2 (95/5) and (b) storage viscosity of PS, PP/PS(70/30) and PS/nano-SiO_2 (95/5) as a function of frequency. Reproduced with permission from [41]. Copyright 2004, Elsevier Science Ltd

3.7 Conclusions

The crystalline morphology and crystalline kinetics of polymer blends which is related to final structural properties, have been considered. In crystallizable polymeric materials, nucleation processes play a leading role. From thermodynamics, in order for nucleation to occur, a minimum critical radius ($r > r^*$) is required. The presence of an amorphous or crystalline (amorphous/crystalline or crystalline/crystalline) component in a binary blend can either increase or decrease the crystallization rate, depending on the effect of the blend composition on the T_g and T_m of the crystallizable component. The rate of crystallization of semicrystalline polymeric materials decreases with increasing concentration of amorphous components as it acts as a diluent. This can be improved by adding nanofillers at low concentrations.

The morphology developed in semicrystalline polymer blends during processing is largely determined by the processing conditions and the type of expulsion of the amorphous component. By controlling shear rate, shear time, and temperature, a morphology transformation from spherulite to shish-kebab could be obtained, leading to increased mechanical properties.

References

[1] Lim, L.-T., Auras, R., Rubino, M. Processing technologies for poly(lactic acid). *Prog. Mater. Sci.* 33, 820–852 (2008)

[2] Spitalsky, Z., Tasis, D., Papagelis, K., Galiotis, C. Carbon nanotube–polymer composites: chemistry, processing, mechanical and electrical properties. *Prog. Mater. Sci.* 35, 357–401 (2010)

[3] Zhang, C., Wang, B., Yang, J., Ding, D., Yan, X., Zheng, G., Dai, K., Liu, C., G, Z. Synergies among the self-assembled β-nucleating agent and the sheared isotactic polypropylene matrix. *Polymer* 60, 40–49 (2015)

[4] Zhang, R.-C., Xu, Y., Lu, A., Cheng, K., Huang, Y., Li, Z-M. Shear-induced crystallization of poly(phenylene sulfide). *Polymer* 49, 2604–2613 (2008)

[5] Wang, J., Yang, J., Deng, L., Fang, H., Zhang, Y., Wang, Z. More dominant shear flow effect assisted by added carbon nanotubes on crystallization kinetics of isotactic polypropylene in nanocomposites. *ACS Appl. Mater. Interfaces* 7, 1364–1375 (2015)

[6] Fang, H., Zhang, Y., Bai, J., Wang, Z. Shear-induced nucleation and morphological evolution for bimodal long chain branched polylactide. *Macromolecules* 46, 6555–6565 (2013)

[7] Muthukumar, M. Communication: Theory of melt-memory in polymer crystallization. *J. Chem. Phys.* 145, 031105 (2016)

[8] Finney, E.E., Finke, R.G. Nanocluster nucleation and growth kinetic and mechanistic studies: A review emphasizing transition-metal nanoclusters. *J. Colloid Interface Sci.* 317, 351–374 (2008)

[9] Taqieddin, A., Allshouse, M.R., Alshawabkeh, A.N. Critical review – Mathematical formulations of electrochemically gas-evolving systems. *J. Electrochem. Soc.* 165, E694 (2018)

[10] Turnbull, D., Fisher, J.C. Rate of nucleation in condensed systems. *J. Chem. Phys.* 17, 71–73 (1949)

[11] Lauritzen Jr, J. I., Hoffman, J. D. Extension of theory of growth of chain-folded polymer crystals to large undercoolings. *J. Appl. Phys.* 44, 4340–4352 (1973)

[12] Song, J., Lu, C., Xu, D., Ni, Y., Liu, Y., Xu, Z., Liu, J. The effect of lanthanum oxide (La_2O_3) on the structure and crystallization of poly(vinylidene fluoride). *Polym. Int.* 59, 954–960 (2010)

[13] Kar, G. P., Bose, S. Nucleation barrier, growth kinetics in ternary polymer blend filled with preferentially distributed carbon nanotubes. *Polymer* 128, 229–241 (2017)

[14] Fillon, B., Lotz, B., Thierry, A., Wittmann, J. Self-nucleation and enhanced nucleation of polymers. Definition of a convenient calorimetric "efficiency scale" and evaluation of nucleating additives in isotactic polypropylene (α phase). *J. Polym. Sci., Part B: Polym. Phys.* 31, 1395–1405 (1993)

[15] Sun, S., Xu, X., Yang, H., Zhang, H. Toughening of poly(butylene terephthalate) with epoxy-functionalized acrylonitrile-butadiene-styrene. *Polymer* 46, 7632-7643 (2005)

[16] Hage, E., Ferreira, L., Manrich, S., Pessan, L. Crystallization behavior of PBT/ABS polymer blends. *J. Appl. Polym. Sci.* 71, 423–430 (1999)

[17] Avrami, M. Granulation, phase change, and microstructure kinetics of phase change. III. *J. Chem. Phys.* 9, 177–184 (1941)

[18] Avrami, M. Kinetics of phase change. II Transformation-time relations for random distribution of nuclei. *J. Chem. Phys.* 8, 212–224 (1940)

[19] Avrami, M. Kinetics of phase change. I General theory. *J. Chem. Phys.* 7, 1103–1112 (1939)

[20] Sencadas, V., Costa, C. M., Gómez Ribelles, J., Lanceros-Méndez, S. Isothermal crystallization kinetics of poly(vinylidene fluoride) in the α-phase in the scope of the Avrami equation. *J. Mater. Sci.* 45, 1328–1335 (2010)

[21] Manna, S., Batabyal, S. K., Nandi, A. K. Preparation and characterization of silver–poly(vinylidene fluoride) nanocomposites: formation of piezoelectric polymorph of poly(vinylidene fluoride). *J. Phys. Chem. B* 110, 12318–12326 (2006)

[22] Ye, H. J., Shao, W. Z., Zhen, L. Crystallization kinetics and phase transformation of poly(vinylidene fluoride) films incorporated with functionalized $BaTiO_3$ nanoparticles. *J. Appl. Polym. Sci.* 129, 2940–2949 (2013)

[23] Achilias, D. S., Bikiaris, D. N., Papastergiadis, E., Giliopoulos, D., Papageorgiou, G. Z. Characterization and crystallization kinetics of in situ prepared poly(propylene terephthalate)/SiO_2 nanocomposites. *Macromol. Chem. Phys.* 211, 66–79 (2010)

[24] Barrau, S., Vanmansart, C., Moreau, M., Addad, A., Stoclet, G., Lefebvre, J.-M., Seguela, R., Crystallization behavior of carbon nanotube–polylactide nanocomposites. *Macromolecules* 44, 6496-6502 (2011)

[25] Chiu, H.-J. Spherulitic morphology and crystallization kinetics of poly(vinylidene fluoride)/poly(vinyl acetate) blends. *J. Polym. Res.* 9, 169–174 (2002)

[26] Phillips, P. J., Rensch, G. J., Taylor, K. D. Crystallization studies of poly(ε-caprolactone). I. Morphology and kinetics. *J. Polym. Sci., Part B: Polym. Phys.* 25, 1725–1740 (1987)

[27] Zeng, J.-B., Zhu, Q.-Y., Li, Y.-D., Qiu, Z.-C., Wang, Y.-Z. Unique crystalline/crystalline polymer blends of poly(ethylene succinate) and poly(*p*-dioxanone): miscibility and crystallization behaviors. *J. Phys. Chem. B* 114, 14827–14833 (2010)

[28] Penning, J., St. John Manley, R. Miscible blends of two crystalline polymers. 2. Crystallization kinetics and morphology in blends of poly(vinylidene fluoride) and poly(1,4-butylene adipate). *Macromolecules* 29, 84–90 (1996)

[29] Qiu, Z., Yan, C., Lu, J., Yang, W. Miscible crystalline/crystalline polymer blends of poly(vinylidene fluoride) and poly(butylene succinate-co-butylene adipate): spherulitic morphologies and crystallization kinetics. *Macromolecules* 40, 5047–5053 (2007)

[30] Yang, J., Pan, P., Hua, L., Zhu, B., Dong, T., Inoue, Y. Polymorphic crystallization and phase transition of poly(butylene adipate) in its miscible crystalline/crystalline blend with poly(vinylidene fluoride). *Macromolecules* 43, 8610–8618 (2010)

[31] Gebrekrstos, A., Sharma, M., Madras, G., Bose, S. Critical insights into the effect of shear, shear history, and the concentration of a diluent on the polymorphism in poly(vinylidene fluoride). *Cryst. Growth Des.* 17, 1957–1965 (2017)

[32] Chen, H.-L., Li, L.-J., Lin, T.-L. Formation of segregation morphology in crystalline/amorphous polymer blends: molecular weight effect. *Macromolecules* 31, 2255–2264 (1998)

[33] Wu, D., Lin, D., Zhang, J., Zhou, W., Zhang, M., Zhang, Y., Wang, D., Lin, B. Selective localization of nanofillers: effect on morphology and crystallization of PLA/PCL blends. *Macromol. Chem. Phys.* 212, 613–626 (2011)

[34] Zhang, L., Wan, C., Zhang, Y. Investigation on the multiwalled carbon nanotubes reinforced polyamide 6/polypropylene composites. *Polym. Eng. Sci.* 49, 1909–1917 (2009)

[35] Yousfi, M., Livi, S., Dumas, A., C-L, J., Greenhill-Hooper, M, Duchet-Rumeau, J. Compatibilization of polypropylene/polyamide 6 blends using new synthetic nanosized talc fillers: Morphology, thermal, and mechanical properties. *J. Appl. Polym. Sci.* 131, 40453 (2014)

[36] Zou, H., Wang, K., Zhang, Q., Fu, Q. A change of phase morphology in poly(*p*-phenylene sulfide)/polyamide 66 blends induced by adding multi-walled carbon nanotubes. *Polymer* 47, 7821–7826 (2006)

[37] Paul, D. Interfacial agents ("compatibilizers") for polymer blends. In *Polymer Blends*. Paul, D. R, Newman, S. (Eds.) (1978), Academic Press, New York, Vol. 2, 35–62

[38] Patra, P.K., Jaisingh, A., Goel, V., Kapur, G.S., Nebhani, L. Crystallization kinetics of compatibilized blends of polypropylene and polyethylenimine. *J. Therm. Anal. Calorim.* 147, 6689–6699 (2022)

[39] Patra, P.K., Jaisingh, A., Goel, V., Luthra, P., Kapur, G.S., Nebhani, L. Comprehensive studies on polyethylenimine filled polypropylene and its potential application in carbon dioxide sequestration. *Polym. Eng. Sci.* 59, 2092–2102 (2019)

[40] Zhang, C., Lan, Q., Zhai, T., Nie, S., Luo, J, Yan, W. Melt crystallization behavior and crystalline morphology of polylactide/poly(ε-caprolactone) blends compatibilized by lactide-caprolactone copolymer. *Polymers* 10, 1181 (2018)

[41] Zhang, Q., Yang, H., Fu, Q. Kinetics-controlled compatibilization of immiscible polypropylene/polystyrene blends using nano-SiO_2 particles. *Polymer* 45, 1913–1922 (2004)

4 Theoretical Studies of Phase Separation in Polymer Blends

■ 4.1 Introduction

Polymer blends have been a key focus of polymer-based research progress since the 1960s. Even though polymer blends were earlier identified as being of interest, academic as well as industrial activity in polymeric blends heightened significantly at the end of the 1960s. Though several factors contributed to this increased attention, one important aspect was the advent of a key polymeric blend for engineering based on polystyrene plus poly(2,6-dimethyl-1,4-phenylene oxide) (PPO).

The miscibility of this blend allows for valuable features across the composition collection. A breakthrough at General Electric [1] culminated in the commercialization of a number of polymeric blends with the common tradename Noryl®. Whereas their industrial impact was clear, scholarly attention was also high. Miscibility in polymer blends was thought to be a rather unusual phenomenon at the time. The discovery of miscibility for PS-PPO (in which no apparent synergistic relationship persists) prompted academic research aimed at better grasping the miscibility of polymeric matrices. The alternative cause of the increase in polymeric blend activities was due to the realization that these materials may possess the same relevance as metal alloys. Styrene polymerization in situ with rubber aimed at making polystyrene with high impact resistance, as well as styrene@acrylonitrile polymerization in the presence of rubber targeted at poly(acrylonitrile butadiene styrene) (ABS) production, are key polymer-based products. The technique discussed here (invented in the 1950s) produced far superior characteristics than simple PS/acrylonitrile copolymer mixes with rubber [2]. In the 1960s, rubber blends outperformed any single unblended rubber composition in terms of tire qualities. Poly(vinyl chloride) (PVC)/butadiene-acrylonitrile rubber was among the initial commercialized partially miscible polymeric blends (in the early 1940s), which yielded a product known as lastingly plasticized PVCs [3].

Theoretical investigations of phase separation in polymeric blends are covered in this chapter. Although much of the research activity at the time focused on comprehending the kind of miscibility in the polymeric blend(s), notable progress has likewise been made in creating technologies for the compatibilization of polymeric blends for immiscible as well as incompatible blend entities. The theoretical basis for thermodynamic interactions in polymer blends will be addressed, as will significant improvements in experimental approaches for measuring thermodynamic interactions (e. g., the heat of mixing).

4.2 Flory–Huggins Theory

Gibbs [4] developed the following basic thermodynamic connection for mixtures (Equation 4.1):

$$\Delta G_{\mathrm{m}} = \Delta H_{\mathrm{m}} - T\Delta S_{\mathrm{m}} \tag{4.1}$$

Where ΔG_{m} denotes the free energy of mixing, H_{m} signifies the enthalpy of mixing, while ΔS_{m} is the entropy mixing, and T is the temperature. ΔG_{m} must be less than 0 for blending to take place. It denotes an essential criterion, but it is insufficient because the expression should first of all be satisfied (Equation 4.2):

$$\left(\frac{\partial^2 \Delta G_{\mathrm{m}}}{\partial \phi_i^2}\right)_{T,P} > 0 \tag{4.2}$$

Flory [5] and Huggins [6] devised the rudimentary theory for determining the miscibility of polymeric blends, which is known as the Flory–Huggins hypothesis. The Flory–Huggins expression is named after the equation that resulted from their research (Equation 4.3):

$$\Delta G_{\mathrm{m}} = kTV\left[\frac{\phi_1}{V_1}ln\phi_1 + \frac{\phi_2}{V_2}ln\phi_2\right] + \frac{\phi_1\phi_{2}{}_{12}kTV}{v_r} \tag{4.3}$$

The heat of mixing is required to be negative for miscibility to be achieved (i. e., $\chi_{12} < 0$). Here, V = signifies the whole volume, V_i represents the molecular volume of the components i, ϕ_i stands for the volume ratio of the i components, k denotes the Boltzman's constant, χ_{12} = Flory–Huggins interface component, while v_r is the volume due to interfacial interactions within the blend (like the unit repeat volume), and r represents the volume referenced. This expression was originally developed for polymer-solvent mixes, however, it can also be applied to polymer mixtures with greater molecular masses.

As the combinative mixing entropy decreases, the Flory–Huggins model is employed to effectively decrease solvent-polymer miscibility in comparison with solvent-solvent blends. Within the limit of high molecular weight polymer mixes, the combinative mixing entropy approaches trivial values with polymer-to-polymer blends. With the aim of achieving miscibility, a negative heat of blending (i.e., $\chi_{12} < 0$) must be produced.

The Flory-Huggins free energy of mixing for every lattice site in the Flory-Huggins binary polymeric blends model for polymers A as well as B can be represented as (Equation 4.4):

$$\frac{\Delta G_m}{k_B T} = \frac{\phi \ln \phi}{v_A N_A} + \frac{(1-\phi)\ln(1-\phi)}{v_B N_B} + \frac{\phi(1-\phi)}{v} \chi_{AB}(T,\phi) \tag{4.4}$$

Where v_A and v_B denote the molar volumes of the segment for binary polymeric matrices, v denotes the volume referenced (usually $v = \sqrt{v_A v_B}$, [cm^3/mol]), and χ_{AB} represents Flory-Huggins interfacial components' parameter (has no dimension).

The mixing entropy, as well as the enthalpy, are estimated by tallying alternative arrangements along with interactions along a given lattice in this lattice theory. It has a number of flaws, including the inability to account for changes in volume during blending, nonrandom blending, and obvious polymeric architecture, among others. As a result, explaining the character of genuine blends, the parameter χ must cover all of these other factors, and additionally it is concentration as well as temperature reliant. With respect to composition, the free energy of mixing 2nd derivative ($\Delta G/k_B T$) is, therefore (Equation 4.5) (in mol/cm^3):

$$G'' = \frac{\partial^2 \left(\frac{\Delta G}{k_g T} \right)}{\partial \phi^2} = \frac{1}{\phi_A v_A N_A} + \frac{1}{\phi_B v_B N_B} - 2\frac{{}_{AB}(T)}{v} \tag{4.5}$$

Figure 4.1 shows thermodynamics of an illustrative, symmetric, binary polymer blend with N = 1000, segment volume v = 100 cm^3/mol [7]. The expression requires that χ_{AB} is composition-independent (otherwise, ϕ explicit derivative is added trivially, and χAB is replaced by $-\left(\frac{\partial^2 \left[\phi(1-\phi)_{AB} \right]}{\partial \phi^2} \right) / 2)$. G'' = 0 at the thermodynamic stability limit, given by the spinodal temperature T_s, as depicted in Figure 4.1(b), and thus:

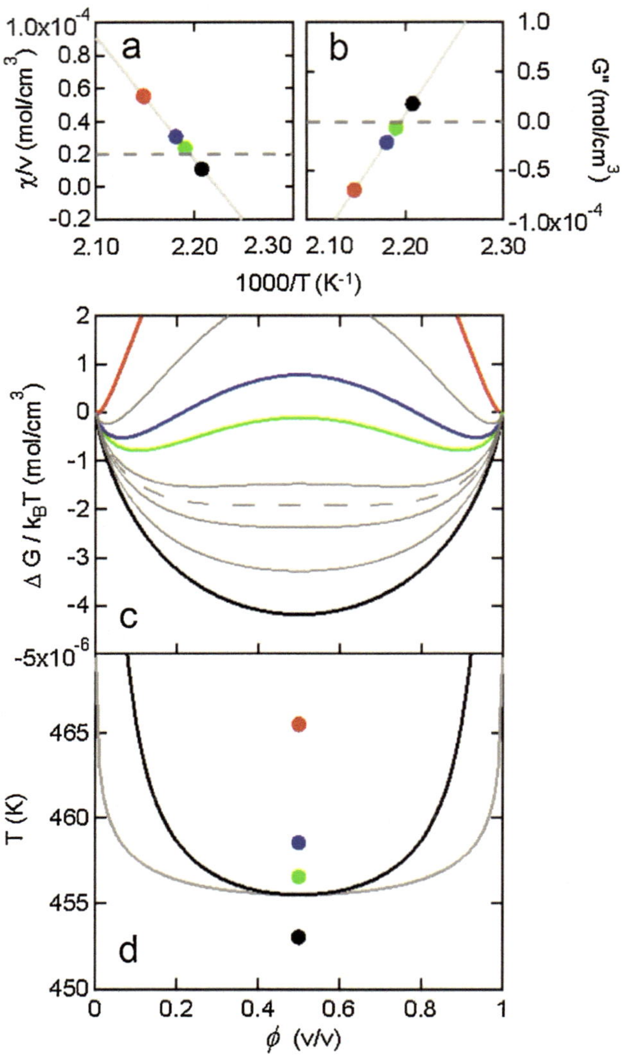

Figure 4.1 Thermodynamics of an illustrative, symmetric, binary polymer blend with N = 1000, segment volume v = 100 cm^3/mol. (a) $\chi_{AB}/v = 0.00166 - \dfrac{0.747}{\text{mol}}/\text{cm}^3$, corresponding, (b) $G''v1/T$ calculated from Equation 4.5, (c) free energy of mixing, $\Delta G_m/k_BT$ computed from Equation 4.4, and (d) Flory-Huggins phase diagram showing the spinodal (black) and binodal (grey) lines. For these parameters $T_c \approx 455.5$ K (or 182.5 °C). The markers correspond to representative points for a 50/50 blend in the one-phase region (black), at $\Delta T \equiv T - T_s = -2.5$ °C and within spinodal region (green, blue, and red) at ΔT = 1.0, 3.0 and 10.0 °C, corresponding to typical temperature quenches. Reproduced with permission from Cabral and Higgins [7]. Copyright 2018, Elsevier Science Ltd

$$\chi_s \equiv \chi_{AB}\left(T - T_s\right) = \frac{v}{2}\left(\frac{1}{\phi_A v_A N_A} + \frac{1}{\phi_B v_B N_B} \right) \tag{4.6}$$

Revising Equation 4.5 and Equation 4.6 results in a facile expression between G" and χ" useable in a make-up independent χ_{AB} as well as monodisperse polymeric materials (If *N* is not present, weight-average *N* > *w* is used instead) (Equation 4.7) (in mol·cm^{-3}):

$$G" = \frac{2}{v}\left(\chi_s - \chi_{AB}(T)\right) \tag{4.7}$$

Figure 4.1(a) and Figure 4.1(b) above show the temperature dependence of χ_{AB}/v and G" for an example blend, as well as the equivalent mixing free energy as well as the phase diagram (Figure 4.1(c) and Figure 4.1(d)). Remember:

$$\frac{\chi}{v} = \left(A + \frac{B}{T}\right)$$

is likewise in [mol/cm^3]. At the spinodal:

$$\chi^N = 2 \Rightarrow \chi = \frac{2}{N}$$

and so

$$\frac{2}{N_v} = \left|A + \frac{B}{T_s}\right) \Rightarrow \frac{B}{T_s} = \frac{2}{N_v} - A \Rightarrow T_s = \frac{B}{\frac{2}{N_v} - A}$$

This theory by Flory–Huggins has been applied by researchers globally in studying polymeric blends and their composites [8–12].

It has been proposed that the thermodynamic theories used to address tricomponent systems differ little from those used to describe binary mixtures [13, 14], which are primarily founded on assessments of the minimums of the total Gibbs energy of a system, which comprises the surface energy, including all interphase barriers. The segregation of the solution is believed to form a core-shell structure with spherical interphase boundaries arising from the process-induced phase separation. Fedoseev et al. proposed a thermodynamic theory of the tricomponent solution (aniline@methylcyclopentane@hexane) with phase separation [13], and they presented the modeling outcomes for many ternary blends having core-shell structures. The computational theory was constructed upon numerical examination of the Gibbs function, which considers the surface energy of the external and internal borders [13]. They revealed that ternary blends having phase-separation conforming to normal state expression, Flory–Huggins solutions, and NRTL solutions had been used as examples [13]. The modeling illustrates thermodynamically steady as well as metastable states that macrosystems lack. The volume and composition of a heterogeneous mixture were demonstrated to determine the area of heterogeneity, the equilibrium formulation of coexisting phases, and the number of metastable conditions. An area of bistability of phase equilibria was uncovered wherein distinct core-shell configurations compete. They discovered that the area of dissimilar states along with aniline within the core phase hinges upon the droplet size from their phase diagram investigation. The coexisting phases comprise cleaner aniline solutions plus hexane in tiny droplets. Such separation occurs due to the evolution of a component with high surface tension (aniline) to a phase with a reduced surface region, lowering the total energy of the system [13].

A two-component derived Flory–Huggins model having a self-consistent analytical expression for binodal intensities has been explored [15]. The authors postulated such versions to widen the efficacy of the Landau–Ginzburg extension outward from the vital point to span the entire subspace, resulting in an exponential scaling law of the dilute phase binodal concentration [15]. Using this method, one can fit investigational protein liquid-liquid phase-separation data and derived variables such as effective chain length and solute-solvent chemical interactions. They estimated the interaction energy effects of distinct amino acids to showcase the power of this concept. Such an analytical framework can also be used to interpret the protein arrangement morphology for liquid-liquid phase separation in a novel way [15].

Pötschke et al. established an equilibrium G^A phase by combining AB as well as BC diblock polymeric matrices with a small quantity of ABC triblock polymeric matrix, which acted as an equilibrium-breaking surfactant [16]. Self-consistent field theory (SCFT) studies showed that the triblock terpolymer removes the potential for the two diblock copolymers to macro-phase split into virtually absolute G phases, hence steadying the G^A stabilization window at diblock stoichiometric combinations. They investigated the feasibility of creating G^A systems by blending diblocks rather than synthesizing G^A-forming triblocks or pentablock copolymers for the first time, proving that the block copolymer mixture approach, which may not demand restrictive triblock designs, will provide an efficient and suitable route to developing optoelectronic materials with thorough photonic band gap web nanostructured materials [16]. They used a relatively large state space for said ternary AB/BC/ABC blend that was meant to achieve rich flow characteristics, including an oscillating gyroid region. They limited their attention to AB as well as BC diblocks of minority block degrees of polymerization $N_A = N_C$ and a familiar value of N_B; thus $N = N_A + N_B = N_B + N_C$ denotes the cumulative degree of polymerization of the diblocks but also $f_A = f_C = 0.357$ represents the volume fraction of their corresponding smaller blocks. Flory–Huggins interaction parameters $\chi_{AB}N = \chi_{BC}N = 20$ as well as $\chi_{AC}N = 35$ were also chosen. The pristine AB, as well as BC diblocks, yielded the G architecture under such conditions. The third part, an ABC triblock polymeric matrix, had such a polymerization degree equal $N' = 1.3428\,N$, having equivalent end blocks, $N_A' = N_C' = 0.4284\,N$, as well as a pivot block with

N_B' = 0.486 N. The SCFT calculations anticipated that the ABC triblock generates alternate lamellae (LA) inside this pure melt state; however, as seen in Figure 4.2, the specific constitution of the triblock is not really crucial. Those polymeric matrices were represented as flexible Gaussian chains having equivalent statistical segment length, "*b*", as well as volume, *v*. Its self-consistent mean-field results obtained by solving mathematically the improved diffusion equation for SCFT and revising the fields for self-consistency had been used to compute the free energies of several candidate phases. The various alternate microdomain morphologies were examined for the mixed systems: alternating gyroid (G^A), alternating diamond (D^A), alternating lamellae (L^A), square-packed alternating cylinders ($C_4{}^A$), as well as hexagonal packed alternating cylinders ($C_6{}^A$). They also included renowned hexagonal cylinder (C_6), twin diamond (D), twin gyroid (G), as well as lamellar (L) phases for the AB as well as BC diblock-rich phases. It also showed the free energy in AB/BC binary mixture system in terms of the AB diblock polymers volume fraction ϕ_{AB}, for each candidate phase. Because of the structural equivalence of ϕ_{AB} as well as ϕ_{BC} diblocks, they claimed that the free energy profiles for AB were symmetric around AB = 0.5. Despite differences in block polymers, which might not produce multiple contemporaneous microphases separated states, these two-component mixes could form both micro- and macrophase separated states. The minority A or C blocks were designed to be thermodynamically irreconcilable, resulting in microphase disjointing into distinct domains split by a universal B blocks matrix within the numerous oscillating phases correlated with binary mixes.

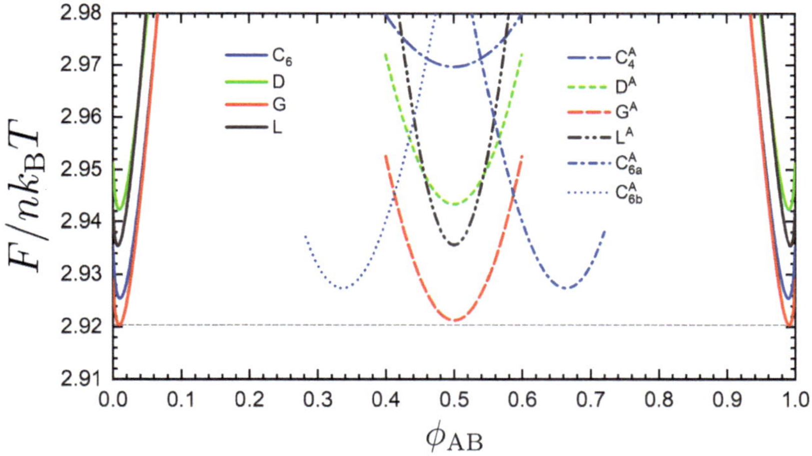

Figure 4.2 Free energy per diblock chain in binary blends of G-forming AB/BC diblocks ($f_A = f_C = 0.357$, $\chi_{AB}N = \chi_{BC}N = 20$, and $\chi_{AC}N = 35$) as a function of AB diblock volume fraction ϕ_{AB} for various alternating phases ($C_4{}^A$, D^A, G^A, L^A, $C_{6a}{}^A$, and $C_{6b}{}^A$) and AB- and BC-rich phases (C_6, D, G, and L). The $C_{6a}{}^A$ and $C_{6b}{}^A$ phases are hexagonally packed alternating cylinder phases with A and B majority cylinders, respectively. The dashed grey tie-line connects the minimums of the AB- and BC-rich G free energy curves. Reproduced with permission from [16]. Copyright 2022, American Chemical Society

As shown in Figure 4.2, combining two G-creating diblocks results in the least free energy in favor of the G^A phase at $\phi_{AB} = 0.5$, so this architecture explicitly outperforms all other oscillating phases recognized. But then again, the texturally symmetric ($\phi_{AB} = 0.5$) G^A the mixture has a marginally greater free energy in comparison to G phases constituted of highly refined AB as well as BC diblock copolymers, implying also that the equilibrium state is macrophase distinction of the two phases. They also speculated that perhaps the well almost deviant free energy of said G^A phase, comparative here to highly refined AB as well as BC G phases, could lead to the development of G^A whenever a relatively uniform solution of the two diblocks is speedily solvent cast, roughly equivalent to the creation of metastable martensite once one mixture of carbon as well as Fe in the context of FCC metal matrix is swiftly chilled to a low temperature.

This is further explained by the fact that diffusion constraints will probably impede the nucleation as well as the evolution of macroscopic G realms, while the transformation to G^A demands no polymeric transfer. Furthermore, the authors offered greater free energy computations for the G-forming diblocks at reduced $\chi_{AB} = \chi_{BC}$ values and discovered that the free energy gap between the mixed G^A phase, as well as the macroscopically partitioned G phases, diminishes at higher segregation strength (χ_N), implying that the solvent casting may become more feasible with higher molecular weight polymeric materials. Figure 4.2 shows a similar free energy plot with supplemental candidate phases (core-shell phases as well as O^{70}), verifying that the outcomes inferred from Figure 4.3 seem to be robust to the incorporation of further matching phases. The authors added modest doses of the ABC triblock polymer to normalize the G^A phase against macrophase secession, hoping that this admixture might well act like an emulsifier by mending the A and C sites inside the interchanging phases. The above strategy is similar to that used to stabilize co-continuous morphological features in homopolymer mixtures of ABC triblock terpolymers. The influence of the ABC triblock ϕ_{ABC} here upon the free energy of the oscillating phases throughout mixtures comprising the same quantity of AB as well as BC diblocks ($\phi_{AB} = \phi_{BC}$) is shown in Figure 4.3. Low doses of the ABC triblock decrease the free energy among all oscillating phases toward a lower limit at ABC $\phi_{ABC} \approx 0.04$; further, than this point, all free energies significantly raise. This same G^A phase has quite a dramatic decrease in free energy ($\approx 0.02nk_BT$), implying that the ternary blend would then stabilize G^A at the equilibrium phase vs. the phase, isolated G regions with essentially pure AB as well as BC deblocks [16].

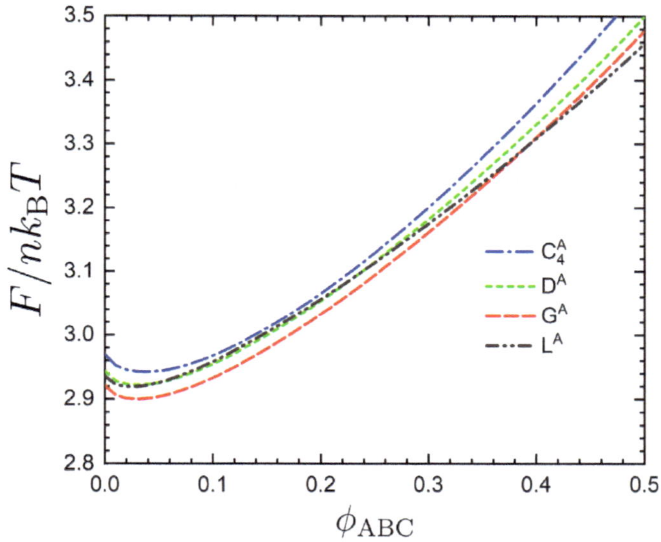

Figure 4.3 Free energy per chain of size N as a function of ABC triblock composition ϕ_{ABC} in the ternary blends with the same amount of the AB and BC diblocks ($\phi_{AB} = \phi_{BC} = (1 - \phi_{ABC})/2$). The diblock parameters and interaction strengths are the same as in Figure 4.2, and the ABC triblock polymer is symmetric ($N^{A'} = N_C' = 0.4284N$ and $N_B' = 0.486N$), where N is the degree of polymerization of the AB or BC diblock. In the neat melt, this ABC triblock forms the L^A phase. Reproduced with permission from [16]. Copyright 2022, American Chemical Society

The ternary-phase schematic of said diblock copolymer rich portion, as well as the corresponding firmness windows crafted by utilizing enormous canonical SCFT solutions, are seen in Figure 4.4.

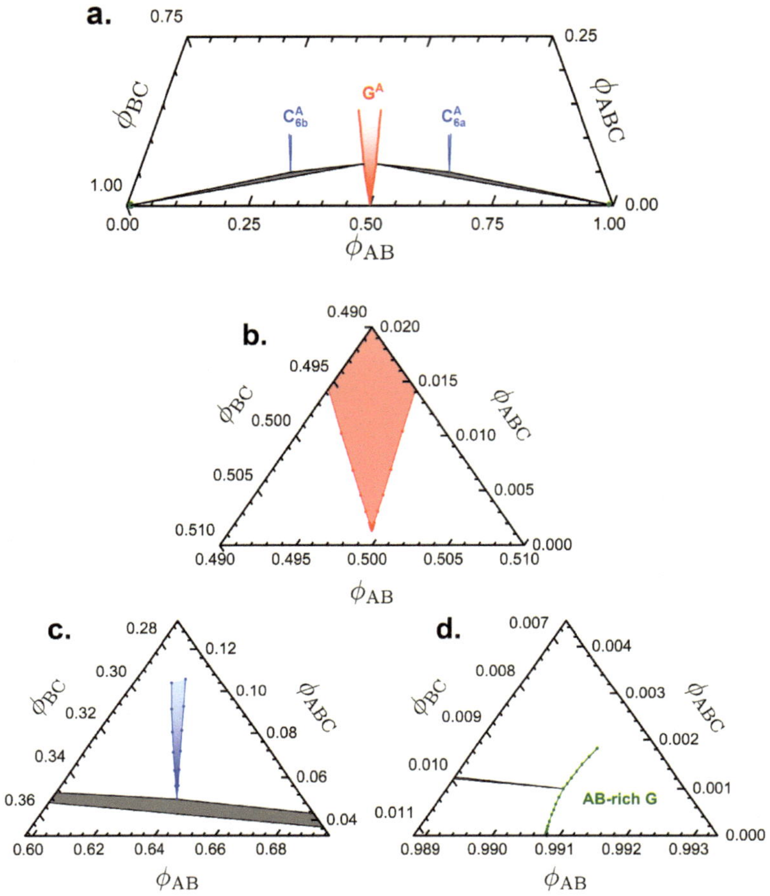

Figure 4.4 Portions of the triangular phase diagram of the AB/BC/ABC ternary blend corresponding to Figure 4.2 and Figure 4.3 for (a) the bottom portion of the ternary-phase diagram, (b) the G^A phase window, (c) the stability window of the $C_{6a}{}^A$ phase with A majority cylinders, and (d) the AB-rich G phase window. The two grey triangles denote the three-phase coexistence windows for the $G^A/C_{6b}{}^A$/BC-rich G phases and the $G^A/C_{6a}{}^A$/AB-rich G phases. Reproduced with copyright permission from [16]. Copyright 2022, American Chemical Society

Intriguingly, the G^A consistency window was revealed to be more or less consistent with the ternary-phase triangle's $\phi_{ABC} = 0$, which further entails that perhaps the least possible quantity of ABC triblocks needed to kick off the G^A the window is unbelievably little as $\phi_{ABC} = 1.32\ 10^{-3}$. If the diblock mixture stoichiometry deviates a little from the desirable G^A condition (i. e., in the two-phase area distinguishing the G^A and G phases), then a small amount of double gyroid defilement will develop in alignment with most of the G^A phase. The authors investigated the effects of changing the block proportion in the ABC triblock upon the efficacy of the said compatibilizing agent. They also demonstrated that the threshold of triblock necessary to create the G^A is almost indifferent to minor differences in the lengths of the B and end blocks. As a result, for a wide series of ABC constituents, <1% of the triblock could perhaps restrict macrophase separation but also result in a stable G^A phase once introduced into a symmetric blend of AB as well as BC diblocks ($\phi_{AB} = \phi_{BC}$), and indeed the detailed configuration of the ABC triblock is fairly irrelevant in the limit of low ABC operating as a type of dispersant. They also discovered that when the segregation capacity of the dual gyroid-creating diblocks rises, the minimum ABC needed to open the G^A window decreases. Because of such character, the block copolymer mixture approach is enticing for manufacturing different length-scale G^A nanomaterials, which are useful for manufacturing photonic band gaps as well as other materials.

It is understandable that even at the intense threshold of a diminished B-block length, i. e., AC diblock polymers, these and other additives would not be efficient in steadying the G^A phase owing to undesirable interactions with the matrix B blocks as well as a high stretching sanction to overcome the A as well as the C network domains. The ternary mix macroscopically segregates into 3 diblock-rich organized phases at this limit. As a result, the middle structures of ABC triblock polymeric matrices would not be too short; given that the mid-block is sufficiently lengthy to bridge the two network areas, macrophase partitioning is repressed at low ABC doses, regardless of the elaborate block length percentages, because connecting is unaffected by the smaller block size. These authors anticipated that the topography of the ternary-phase diagram at significant amounts of the ABC triblock polymer matrix would be susceptible to their architecture. A G^A forming ABC triblock polymer, for instance, will most probably result in a G^A a window that encompasses the entire ternary-phase illustration that will be researched further in a research investigation.

While their main objective was to stabilize the G^A phase, they hypothesized that exist numerous alternating cylinder phases of different configurations or C-cylinders that might form stable phases at different hybrid structures, like compositions in metal-alloys at low values of ϕ_{ABC}. For simplicity, let us designate such hexagonally packed solid solution of cylinders as $C_{6a}{}^{SS}$, where the majority of the tubes are AB-rich, but the AB- and BC-rich cylinders are scattered arbitrarily upon that hexagonal structure. Their calculations on two different $C_{6a}{}^{SS}$ phases demonstrated that they were not as effective as the $C_{6a}{}^{A}$ phase at low triblock concentrations. They concentrate on the phase diagram's low ϕ_{ABC} regime, where the G^A phase is steadied via the "surfactant-kind" activity of the ABC triblock. Other ordered phases that are not examined here may develop at medium-to-high triblock concentrations. Examining all conceivable phase stability windows might entail probing the entire triangular phase diagram. Throughout this perspective, it would be important to observe the stability of previously mentioned candidate phases, such as distinct binary cylinder phases with uneven coordination numbers as well as sphere-network hybrid architectures, which could be the topic of future research. While solid-solution-like behavior may complicate the cylinder-forming zone of the phase diagram, the G^A the region is easier. Unlike the alternating cylinder as well as alternated sphere phases, which may have distinct one or zero-dimensional domains, respectively, alternating system phases do not have a solid solution owing to the 3D interconnection domain geometry [16]. As a result, the G^A consistency window is located in the ternary-phase diagram at a precise stoichiometric proportion of polymeric mix compositions, analogous to line compounds or else phase fields in metal alloys. As a result, the ABC triblock polymeric materials operate as a form of surfactant, breaking the symmetry of the G phases in the formation of the G^A architecture [16].

A team of scientists reportedly used Flory–Huggins theory as well as the Hansen solubility parameters to generate theoretic binodal lines for ternary as well as quaternary blends, which were confirmed by cloud titration tests. They discovered that adding GO to PES@PVDF blended casting solutions upsurges the inherent thermodynamic volatility, resulting in slimmer selective top layers and bigger porosities until the high viscosity of the GO-containing solutions inhibits solvent–nonsolvent demixing [14]. Furthermore, the Flory–Huggins statement is factual in forecasting the critical point position in ternary systems and binodal line mobility just after the inclusion of GO into PES and PVDF of various molecular weights. PES/GO displayed a lower binodal line shift compared to PVDF/GO, resulting in stronger thermodynamic instability, a faster solvent/nonsolvent demixing pace, and the creation of higher microporosity within the membrane [14].

Purohit and Taylor, in their study, utilized the Flory–Huggins model to investigate drug-to-polymer miscibility [12]. The phase behavior of ITZ@HPMC membranes produced using solvent evaporation was investigated before as well as after heating. The traditional technique for evaluating miscibility, DSC, was combined with incipient analytical techniques like fluorescence imaging, fluorescence spectroscopy, AFM coupled with nanoscale IR spectroscopy, and nanothermal analysis (AFM-nanoIR-nanoTA). Their DSC data revealed a single T_g process in systems having 10% to 50% drug loading, indicating that the ASDs were miscible, although phase separation was seen in all the membranes using the other characterization methods. The phase separation happened at the submicron scale, according to the AFM-coupled approaches. When the membranes were heated, it was discovered that the ASD components were mixed [12]. The findings shed fresh light on the phase performance of itraconazole@HPMC colloid but also imply that the emerging analytical approaches addressed were promising for determining miscibility as well as microstructure in drug-to-polymer systems [12]. So, these authors have shown that the Flory–Huggins model can also be used in drug-polymer systems.

Flory–Huggins phase diagrams have also been explored as a tool for pre-fabrication in the production of amorphous-solid dispersions with a comparison of hot-melt extrusion to spray drying [17]. In another instance, Tian et al. [18] used the

Flory-Huggins interaction model to develop drug@polymer thermodynamic contour plots and evaluated the importance of temperature as well as drug weight percentage in solid dispersions.

■ 4.3 Chad–Hilliard–Cook Theory

This section analyzes the phase separation fundamentals of the Cahn–Hilliard–Cook (CHC) model, which, in recent years, has been widely utilized to explain dynamic scattering experimentations on twofold polymeric blend(s) as well as connecting its application to multicomponent combinations. The nonlinear Langevin equation that the CHC-theory is based on is as follows (Equation 4.8):

$$\frac{\partial\phi(q,t)}{\partial t}=-\beta(q)\mu(q)+(q,t) \tag{4.8}$$

Where $\phi(q,t)$ denotes the Fourier transform of one of the binary mixture's incompressible components' local volume fraction $\phi(r,t)$, $\Lambda(q)$ represents the q-dependent Onsager coefficient, $\mu(q,t)$ stands for the Fourier transform local chemical potential difference $\mu(r,t)$, while $\eta(q,t)$ is the random force cased by thermal fluctuations. The latter is considered to be a fixed white noise process having zero mean as well as auto-covariance, which is calculated using the fluctuation-dissipation theorem:

$$(q,t)n(-q,t')=(q)q^2\delta(t-t')$$

Van Kampen [19] questioned the use of the Langevin equation approach to representing fluctuations in nonlinear macroscopic systems, as done in the CHC theory, for reasons will be discussed here. With the aid of van Kampen's scheme size extension, Akcasu [20] presented the correct approach to the use of Langevin model technique for nonlinear systems in 1977.

There has been great application of this theory by researchers for the study of polymer blends [7, 11, 21]. Karim et al. [11] showed in their simulations that the presence of filler particles in a phase separation system resulted in the creation of concentration waves having the symmetry of the filler particles. AFM analysis on ultrathin polystyrene plus poly(vinyl methyl ether) (PVME) blended matrix films with a low concentrations (wt.%) of the modelled nanofiller particles (nSiO_2 particles; diameter of 100 nm) was used to investigate their predictions. The filled blend films were spun-cast on acid-cleaned SiO_2 wafers, and phase separation was induced by a temperature jump into the two-phase region (T ~ 145 °C) of the bulk polymer blend. It was demonstrated that the SiO_2 nanoparticles were linked to the substrate by washing the polymeric membrane with solvent and that the nanoparticles represented a quenched disorder agitation of the film phase separation. During the intermediate stage of phase separation, the presence of the SiO_2 nanoparticles caused the progression of spherical composition waves around the SiO_2 nanoparticles. Also, as the circumstantial spinodal phase-separation pattern grew far larger than the reinforcing nanoparticles, these target patterns disintegrated. Their findings were in line with ideal two-dimensional CHC simulations of phase separation in polymeric mixtures with low filler particle concentrations [11].

Henderson and Clarke [21] used the Cahn–Hilliard theory model a two-step quench method for symmetric binary polymer mixtures. The blend was quenched from a stable state in the one-phase region to an unstable state in the two-phase region where it undergoes spinodal decomposition. The mixture was allowed to coarsen for two different time periods before a second quench was applied to a point further inside the unstable region, as shown in Figure 4.5 and Figure 4.6. Lesser secondary areas emerged temporarily in the principal areas gained from the first quench phase during the preliminary stages following the second quench. The resulting peak from the second quenched stage was separated from the structured component; the linearized theory was also used to analyze the initial growth phases. The greatest degree of the resulting phase separation was calculated as a quench depth function, a finding they believe will spur more research [21].

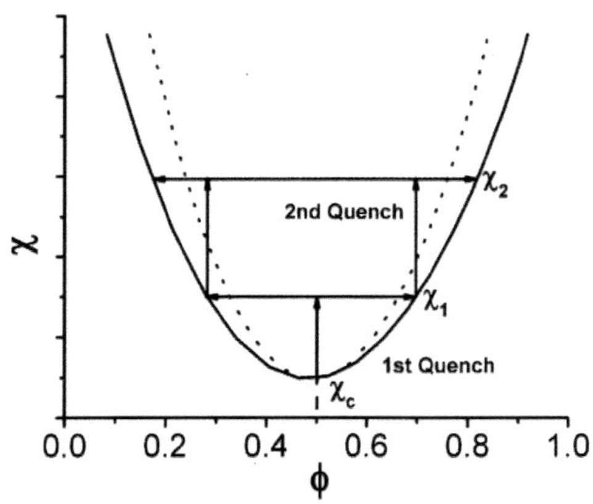

Figure 4.5 Phase diagram illustrating the two-step quench phase separation process in a symmetric binary polymer blend. Reproduced with permission from [21]. Copyright 2018, American Chemical Society

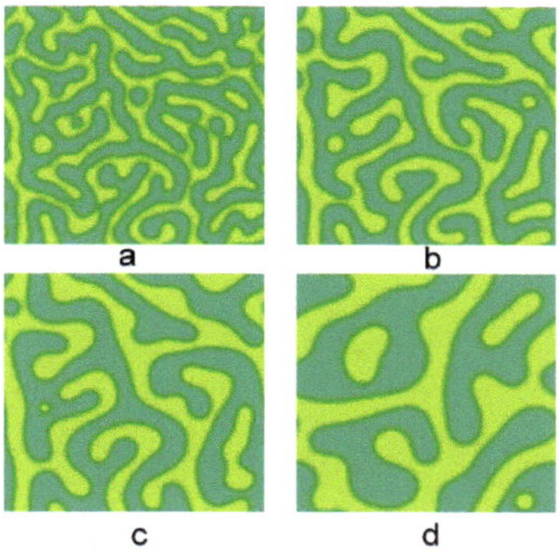

Figure 4.6 Development of the morphology following a quench into the two-phase region with χ_1 = 0.0104. Quench times after the quench are τ_1 = (a) 625, (b) 1250, (c) 2500, and (d) 7500. The initial conditions (random noise) were different for (c) and (d); hence, the structure in (d) is not simply a coarsened version of the structure in (c). Reproduced with permission from [21]. Copyright 2018, American Chemical Society

Snyder, Meakin, and Reich [22] used conventional light scattering techniques to investigate the development time of the structural system in PS/PVME merges. They showed that by using the light scattering invariant and the angular dependence of scattering, the initial phases of phase-separation in the structure of the linearized CHC model can be quantitatively described [22]. It is also suggested that the procedure could be used to determine the spinodal curve by extrapolating the computed diffusion constant to the point it shifts sign, as shown in Figure 4.7.

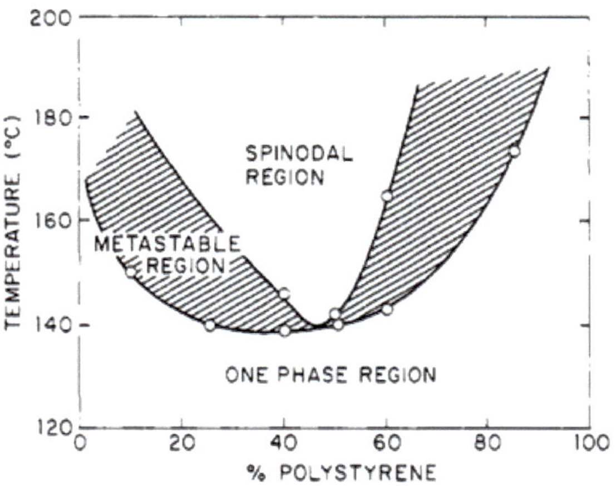

Figure 4.7 Phase diagram determined for the PS/PVME system. The binodal was determined by using cloud point measurements and the behavior of the glass transition. The spinodal curve was determined by extrapolating Reprinted with copyright permission from [22]. Copyright 1983, American Chemical Society

The phase separation free energy was described with the aid of the Flory–Huggins theory, further revised by Gennes, by considering the molar free energy of the polymer blends, thus (Equation 4.9) [22]:

$$f = RT\left(\frac{\phi_1}{v_1}\ln\phi_1 + \frac{\phi_2}{v_2}\ln\phi_2 - \frac{\phi_1\phi_2\chi}{v_1} + \frac{a^2}{36\phi_2\left(1-\phi_2\right)}\left(\nabla\phi_2\right)^2\right) \tag{4.9}$$

Where ϕ_1 and ϕ_2 denote the concentration of the polymers 1 and 2 with molar volumes v_1 and v_2, the mean-field interaction is represented by ($\chi = A + BV/RT$) where A and B are the constants assumed commonly as a parameter dependent on the temperature T, while a^2 is the end-to-end expanse mean square [22].

4.4 Landau–Ginzburg Theory

The Landau-Ginzburg theory is the foundation theory for understanding phase transition in general. The Landau-Ginzburg theory has been applied to a variety of concepts, including liquid crystal transitions as well as superconductivity, solid-solid and liquid-solid phase transitions along with superfluid helium transitions, ferromagnetic, paramagnetic phase transitions, and polymer system phase transitions. The Landau-Ginzburg theory examines the relationship between mesoscopic and microscopic parameters and the characteristics of the system. Leibler studied phase transitions in diblock copolymers [23].

Phase transitions are generally the result of particle interaction, with the Bose–Einstein condensation as an example, while cooperative phenomena are also featured. The Landau-Ginzburg theory is the foundation theory for understanding phase transition in general. The nonlinear diffusion theory developed for large, single-phase structures can predict phase separation though not the separation kinetics or morphology advancement. To address this inadequacy, the Landau–Ginzburg theory introduces the concept of gradient energy.

The concepts of a least possible bifurcation dimension as well as the meticulously associated least possible droplet dimension within a two-phase blend are derived from this functional minimization to acquire equilibrium profiles in miniature systems. The Landau-Ginzburg theory also results in simplified chemical potential, driving diffusion in miniature as well as larger systems at a moderate analysis.

This section aims to look at how the Landau–Ginzburg model can be (has been) applied to polymeric systems. Regardless of its evident constraints (applicable only for inadequately first order as well as uninterrupted phase shifts), this approach offers a wealth of information about polymeric systems like diblock copolymers and their blends, as well as crystalline polymers in their liquid form. The factors distinguishing the Landau-Ginzburg free energy in polymeric systems could be associated with the mesoscopic parameters of the polymer systems. For example, the mesoscopic parameters that characterize interactions are related to the microscopic interaction parameters via the integrals of the direct correlation functions. As a result, the Landau-Ginzburg model results shed light on the relationship between the mesoscopic and microscopic factors of the system and its characteristics. The A homopolymer and B homopolymer blend is the most basic polymer mixture. The mean-field Flory-Huggins free energy as well as de Gennes random phase approximation (RPA) for scattering intensity are used in a very productive technique for studying the system close to its consulate (crucial) point: both can be expressed using the Landau-Ginzburg model. RPA and the Flory-Huggins models are compatible with two mean-field theories. They are called mean-field because they consider only the most likely polymer chain structures and, more significantly, ignore critical long-wavelength fluctuations. De Gennes [24] first proposed the possibility of the effective application of the mean-field theory for high-molecular-mass polymeric admixtures, as opposed to admixtures having low molecular mass, where the mean-field theory fails near the critical point. With respect to the real sense, the mean-field theory is exact in the limit of $N_A, N_B \to \infty$. The mean-field approach is quantitatively correct if the correlation length $\sim \left(\frac{T}{T_c} - 1\right)^{-1}$ is comparable to, or smaller than, the typical length scale of the system, i.e., the radius of gyration $\sim \sqrt{N}$. When two length scales for large N are compared, we find that the mean-field theory fails remarkably close to the critical point, i.e., for large N (Equation 4.10).

$$\frac{|T - T_c|}{T_c} \sim 1/N \tag{4.10}$$

This is known as the Ginzburg criterion. The Ginzburg region is the area about the critical point wherever the mean-field model fails. The critical temperature shift, T_c, in regard to T_c^{mf}, in regard to (as computed by Flory-Huggins theory) is (Equation 4.11):

$$\frac{|T_c - T_c^{mf}|}{T_c^{mf}} \sim 1/\sqrt{N} \tag{4.11}$$

In the one-loop approximation, the shift caused by fluctuations is obtained: being greater than the entire Ginzburg zone. Equation 4.10 above is derived from the analysis of scattering intensity, $I(q)$, wherein q is the scattering wavevector. For $T \to T_c$ as well as $q \to 0$, we can assume:

$$I^{-1}(0) \sim (T/T_c - 1)^{-\gamma}$$

Where $\gamma = 1$ within the mean-field theory. This relation breakdown is commonly used in experiments to calculate the Ginzburg region approximately [25], and thus Ginzburg region computations are based on the scattering intensity [26]. When you get close enough to T_c, fluctuations become significant and change from 1 to its common value around 1.26. This difference could be isolated by plotting:

$$\log I(0) \text{ vs. } \log\left|\frac{T}{T_c} - 1\right|$$

Diverse polymeric matrices can be amalgamated into a single material in a variety of approaches aimed at broad advanced applications. Many potential applications are dependent on the morphology of the system, which is dependent upon this single-chain behavior.

There has been great use of the Landau-Ginzburg theory in polymeric blends [26–28]. For example, Nauman et al. investigated reaction-induced phase separation (RIPS) in epoxy blends and they found that the toughness increase over pure epoxy resins was determined by the phase morphology: domain size, volume ratio, as well as phase architecture [29]. The Flory-Huggins thermodynamics, Landau-Ginzburg functional, as well as a modified Cahn-Hilliard diffusion theory were used to examine a compositional cooling process for integrating incompatible polymeric matrices in which two or more polymeric matrices are dissolved in a co-solvent which was then removed by flash devolatilization [29]. When one of the blend phases is small, a uniform size dispersion will take place, while in a situation where the volume of the phases is commensurate, a continuous interpenetrating network will occur.

Burghardt reported the linear viscoelastic outcomes for a polymer-based bicontinuous microemulsion, where experiments made feasible by the relatively elevated viscosities of the main components were explored [30]. After deducting the viscous background influence from the absolute constituents, the studied microemulsion exhibited excess viscoelastic attributes that resemble the Rouse model for polymer dynamics. The results were compared to Pätzold and Dawson's time-dependent Landau-Ginzburg model, which uses structural factors obtained via neutron scattering as input [30]: this theory was also shown to accurately depict the crucial attributes of viscoelastic attributes of the blend.

The adoption of the Landau-Ginzburg theory to study the dielectric properties of thiourea-based composites was investigated by Stukova et al. [31]. Their study found, with the help of Landau-Ginzburg theory, that introducing polar particles ($BaTiO_3$, $PbTiO_3$, or $LiNbO_3$) into $SC(NH_2)_2$ increased the dielectric permittivity as well as promoted diffusion phase transition. The greatest influence on the properties of thiourea was exerted by barium titanate particles, which was evident by a decrease in ferroelectric phase separation temperature as well as an increase in temperature hysteresis. The magnitude of the effect was defined by the dielectric permittivity of the added particulate [31].

Landau–Ginzburg theory, as well as the other earlier mentioned theories used in the field of material science, have been of immense help for decades. However, it is necessary to further upgrade these models.

■ 4.5 Conclusions

Polymer blends have undoubtedly been amongst the most important study niche in polymer science and polymer technology within recent decades, from both academic and industrial standpoints. Several academics and industry companies, including Union Carbide and General Electric, as well as the Dutch team of Koningsveld and Kleindienst, deserve credit. Several endeavors of others, as well as collaborative initiatives, have had a significant impact. Polymer mixtures have been the subject of numerous books. Diablo's vision of several polymer models has supported researchers in studying different blends, whether miscible or immiscible. These blends have been applied both at an industrial or laboratory scale to fabricate diverse materials for application in advanced material science and technology.

It has been established that the CHC theory having FHG free energy is unsuitable for describing the ending stages of phase-separation, exclusive of taking hydrodynamic flows into account [32]. It was later proven via experimentations that at what time the bicontinuous architecture breaks into drops towards the last stages of phase-separation (which would occur as the result of either the discrete size effect or perhaps the wettability of the surrounding walls), the system develops as per coalescence-triggered coalescence phenomenon [32]. Any mechanism of domain coalescence in this mechanism triggers hydrodynamic flow, which pushes additional domains towards themselves as well as causes additional coalescence events. This "chain-reaction process" is said to be self-sustaining and eventually comes to an end. Stopping the mechanism, interestingly, halts the system's morphology within a certain local minimum of the Gibbs free energy. Additional separation could occur if only as well as only if an external field (e.g., gravitational or electrical field) is employed, causing the system to exit the local minimum. As a result, the final phase in the phase-separation kinetics, which also leads to the total separation of the system into two components with a unified interface in both, requires the use of an external field combined with the factors that emerge all through the phase transition, like density, concentration, as well as the local dielectric constant [32].

References

[1] Cizek, E. P. Blend of a polyphenylene ether and a styrene resin. U. S. Patent 3 383 435 (1968)

[2] Henton, D. E., Bubeck, R. A. The manufacture and physical properties of rubber-toughened styrenics. In *Polymer Toughening*. Arends, C. B. (Ed.) (1996), CRC Press, Boca Raton, 237–292

[3] Henderson, D. E. Mixture of polymerized materials. U. S. Patent 2 330 353 (1943)

[4] Gibbs, J. W. A method of geometrical representation of the thermodynamic properties by means of surfaces. *Trans. Conn. Acad. Arts Sci.* 2, 382–404 (1873)

[5] Flory, P. J. Thermodynamics of high polymer solutions. *J. Chem. Phys.* 9, 660–660 (1941)

[6] Huggins, M. L. Some properties of solutions of long-chain compounds. *J. Phys. Chem. A* 46, 151–158 (1942)

[7] Cabral, J. T., Higgins, J. S. Spinodal nanostructures in polymer blends: On the validity of the Cahn-Hilliard length scale prediction. *Prog. Polym. Sci.* 81, 1–21 (2018)

[8] Pascault, J. P. In *Macromolecular Symposia*. Vol. 93, 43–51 (Wiley Online Library, 1995)

[9] Muzata, T. S., Jagadeshvaran, P., Bose, S. Nanoparticles influence miscibility in LCST polymer blends: from fundamental perspective to current applications. *Phys. Chem. Chem. Phys.* 22, 20167–20188 (2020)

[10] Xiao, T., Wang, P., Yang, X., Cai, X., Lu, J. Fabrication and characterization of novel asymmetric polyvinylidene fluoride (PVDF) membranes by the nonsolvent thermally induced phase separation (NTIPS) method for membrane distillation applications. *J. Membr. Sci.* 489, 160–174 (2015)

[11] Karim, A., Douglas, J. F., Nisato, G., Liu, D.-W., Amis, E. J. Transient target patterns in phase separating filled polymer blends. *Macromolecules* 32, 5917–5924 (1999)

[12] Purohit, H. S., Taylor, L. S. Miscibility of itraconazole–hydroxypropyl methylcellulose blends: Insights with high resolution analytical methodologies. *Mol. Pharmaceutics* 12, 4542–4553 (2015)

[13] Fedoseev, V. Size Effect in the Phase Separation in Three-Component Solutions. *Russ. J. Phys. Chem. A* 96, 813–817 (2022)

[14] Mohsenpour, S., Leaper, S., Shokri, J., Alberto, M., Gorgojo, P. Effect of graphene oxide in the formation of polymeric asymmetric membranes via phase inversion. *J. Membr. Sci.* 641, 119924 (2022)

[15] Qian, D., Michaels, T. C., Knowles, T. P. Self-consistent solution to the Flory–Huggins model and applications in protein liquid-liquid phase separation. *J. Phys. Chem. Lett.* 13, 7853–7860 (2022)

[16] Park, S. J., Bates, F. S., Dorfman, K. D. Alternating gyroid in block polymer blends. *ACS Macro Lett.* 11, 643–650 (2022)

[17] Tian, Y., Caron, V., Jones, D. S., Healy, A.-M., Andrews, G. P. Using Flory–Huggins phase diagrams as a pre-formulation tool for the production of amorphous solid dispersions: a comparison between hot-melt extrusion and spray drying. *J. Pharm. Pharmacol.* 66, 256–274 (2014)

[18] Tian, Y. et al. Construction of drug–polymer thermodynamic phase diagrams using Flory–Huggins interaction theory: identifying the relevance of temperature and drug weight fraction to phase separation within solid dispersions. *Mol. Pharmaceutics* 10, 236–248 (2013)

[19] Quack, M. Detailed symmetry selection rules for chemical reactions. In *Symmetries and Properties of Non-Rigid Molecules: A Comprehensive Survey*. Maruani, J., Serre, J. (Eds.) (1983), Elsevier, 1983

[20] Akcasu, Z. A. Fluctuations of macrovariables in nonlinear systems: Langevin equation approach. *J. Stat. Phys.* 16, 33–58 (1977)

[21] Henderson, I. C., Clarke, N. Two-step phase separation in polymer blends. *Macromolecules* 37, 1952–1959 (2004)

[22] Snyder, H. L., Meakin, P., Reich, S. Dynamical aspects of phase separation in polymer blends. *Macromolecules* 16, 757–762 (1983)

[23] Leibler, L. Theory of microphase separation in block copolymers. *Macromolecules* 13, 1602–1617 (1980)

[24] De Gennes, P. Qualitative features of polymer demixtion. *J. Phys., Lett.* 38, 441–443 (1977)

[25] Meier, G., Schwahn, D., Mortensen, K., Janssen, S. On the crossover from Ising to mean-field behaviour in compatible binary-polymer blends. *Europhys. Lett.* 22, 577 (1993)

[26] Hołyst, R., Vilgis, T.A. Critical temperature and concentration versus molecular weight in polymer blends: Conformations, fluctuations, and the Ginzburg criterion. *J. Chem. Phys.* 99, 4835–4844 (1993)

[27] Hołyst, R., Vilgis, T.A. The structure and phase transitions in polymer blends, diblock copolymers and liquid crystalline polymers: The Landau–Ginzburg approach. *Macromol. Theory Simul.* 5, 573–643 (1996)

[28] Nauman, E.B., He, D.Q. Nonlinear diffusion and phase separation. *Chem. Eng. Sci.* 56, 1999–2018 (2001)

[29] Nauman, E. et al. Compositional quenching: A process for forming polymer-in-polymer microdispersions and cocontinuous networks. *Chem. Eng. Commun.* 66, 29–55 (1988)

[30] Burghardt, W.R., Krishnan, K., Bates, F.S., Lodge, T.P. Linear viscoelasticity of a polymeric bicontinuous microemulsion. *Macromolecules* 35, 4210–4215 (2002)

[31] Stukova, E.V., Baryshnikov, S.V., Koroleva, E.Y., Milinskiy, A.Y. Dielectric investigation of the composites based on thiourea. *St. Petersburg Polytech. Uni. J.l: Phys. Math.* 1, 9–14 (2015)

[32] Fiałkowski, M., Hołyst, R. Dynamics of phase separation in polymer blends revisited: morphology, spinodal, noise, and nucleation. *Macromol. Theory Simul.* 17, 263–273 (2008)

5 Processing of Phase-Separated Blends

5.1 Introduction

Design and manufacturing procedures are critical because they could influence homogeneous mixing, morphology, as well as, as a result, the final attributes of blends and composites. Polymer blends and composites are commonly made using the processes of melt blending, solution blending, as well as in situ polymerization. Solid-state shear pulverization, latex blending, freeze-drying, as well as coagulation-based spinning are also promising approaches [1].

5.2 Processing Methods

5.2.1 Melt Compounding

Melt mixing is one of the most common ways to make a polymer blend that is both cost-effective and environmentally beneficial. The polymeric matrix/matrices is/are heated to form a melt in this process, while mixing takes place through high-shear forces. Regarding melt mixing, rotating screws can provide a significant amount of mechanical energy for consistent polymer mixing. Reliant upon the rheological as well as thermodynamic qualities of the polymeric components, as well as the processing circumstances, melt blending of two, as well as additional distinct polymeric matrices, results in diverse architecture. The morphology of the produced mixes has a significant impact on the final attributes of the system. A melt mixer (e.g., Brabender mixer, Haake mixer), as well as an extruder, can be used to mix the melt in a batch or continuous operation. Melt processing is thought to be a sustainable choice for preparing polymeric blends for large-scale synthesis in industrial settings with the use of a twin-screw extruder (as depicted in Figure 5.1) for pilot- and industrial-scale processing of thermoplastic polymers at the forefront [1–4].

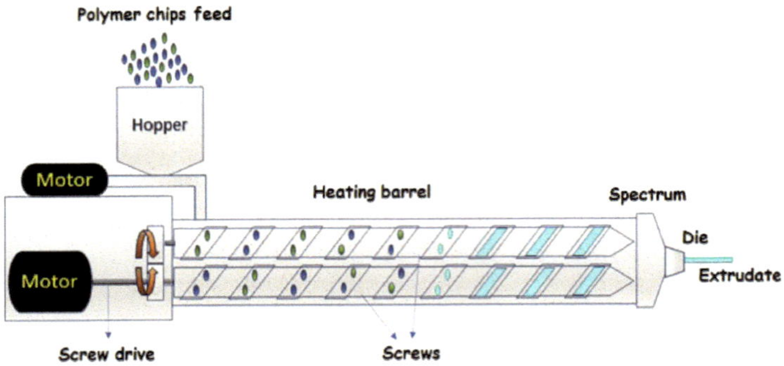

Figure 5.1 Schematic illustration of a typical twin-screw extruder

With the aid of melt blending/compounding, it has been shown that drop coalescence is common in immiscible blend systems, resulting in macroscopic phase separation as well as severe property degradation. Random copolymers are reportedly revealed as being effective in reducing drop coalescence along with boosting the compatibility of immiscible polymeric matrices [5]. It has also been observed in melt compounding of polymeric blends that additives (impact modifiers, nanoparticles, nanofibers, etc.) migrate from one matrix to the other, such as from the minor to the major phase [6] or major to minor phase, or the positioning of the additive at the interface of the blends is common.

Dekkers et al. [6] observed that mechanical peeling, as well as partial solubility of poly(butylene terephthalate) (PBT) in bisphenol A polycarbonate (PC), proved to assist the migration of IM from PBT to PC. They also noticed that several of the core-shell impact modifier (IM) molecules that had migrated into the PC phase initially in the compounding process still were encircled by PBT filaments, indicating that they had been ripped from the softened PBT surface. Even as mixing progressed, those filaments vanished. Their investigations in the microscope hot stage provided some insight into the way this "cleaning" process occurred. The PBT was seen to dissolve in the surrounding PC in samples that were temporarily remelted. These filaments were non-existent when recrystallization, as well as phase separation, was prevented by quick quenching in ice water [6].

The creation of microcapacitor CNTs network in PBS/PVDF polymeric blends via phase separation caused by crystallization towards high dielectric constant "κ" as well as low loss has been postulated to be achievable through a melt compounding approach by Mao et al. [7]. They proposed a novel crystallization approach reliant upon stepwise crystallization to customize the distinguished dispersal as well as distribution of the CNTs in a partially miscible melt mixture of PBS/PVDF via crystallization-triggered phase-separation as well as interfacial engineered inclination amongst CNTs and the polymeric constituents to achieve high dielectric constant as well as the exclusion of PBS low dielectric at the growing anterior of PVDF spherulites, which was also found in the boundary regions of PVDF spherulites throughout the stepwise crystallization course, according to their results. Furthermore, due to the stronger interfacial interface amongst CNTs as well as PBS, CNTs were concentrated within the PBS-abounding area, leading to a high accumulation of the CNTs within PVDF interspherulites [7]. Hence, the dielectric constants of said (nano)composites were enhanced roughly 5–24 times when contrasted to the nanocomposites produced via fast cooling, while the dielectric loss of the nanocomposites remains low [7].

The morphologies produced from polymer@polymer melt compounding are ordered microphases, fibers, co-continuous, double emulsion, drops, and laminar morphologies, as depicted in Figure 5.2 [8].

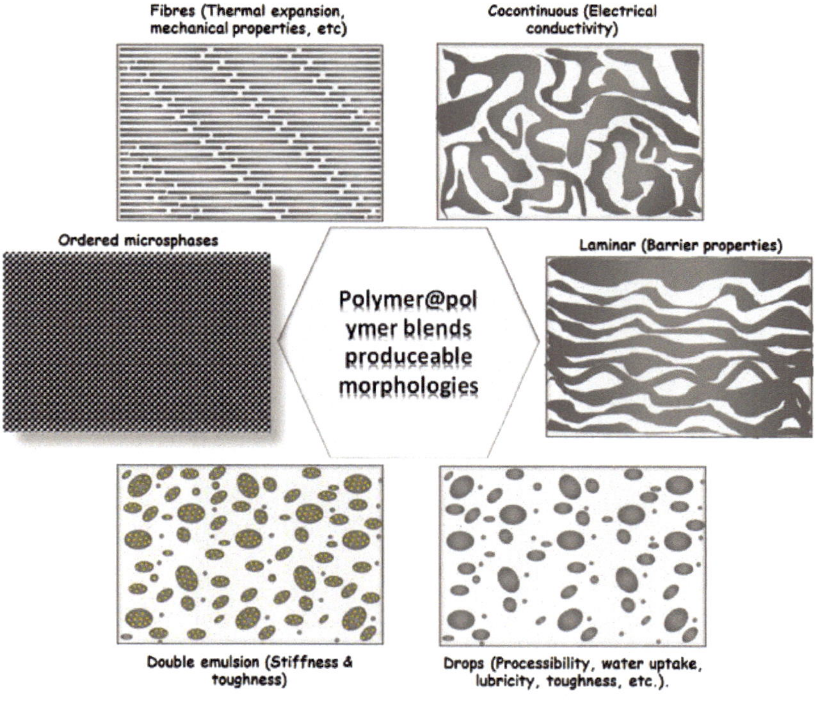

Figure 5.2 Schematic illustration of polymer@polymer blends producible morphologies. Reproduced with permission from [8]. Copyright 2000, Wiley.

These morphologies have also been reported in a research work by Stoclet, Seguela, and Lefebvre in their study of polylactide/polyamide11 blends, as depicted in Figure 5.3 and Figure 5.4 [9]. They discovered that raising the PLA content to 50–60% results in a relatively rough, worn surface with no confirmation of the imprint of phase-separated regions. This topographical difference, which is often created by the liberation of scattered domains from the matrix just at the fracture surface in SEM specimens, appeared to be absent for mixtures within the mid-composition level. This suggests the existence of co-continuous interwoven phases that split up in a fragile way below the T_g. PA11 has a fiber architecture with several globules scattered inside the PLA matrix, with PLA percentages ranging between 60–80%. The fiber width remained substantially more than the diameter of the typical granules.

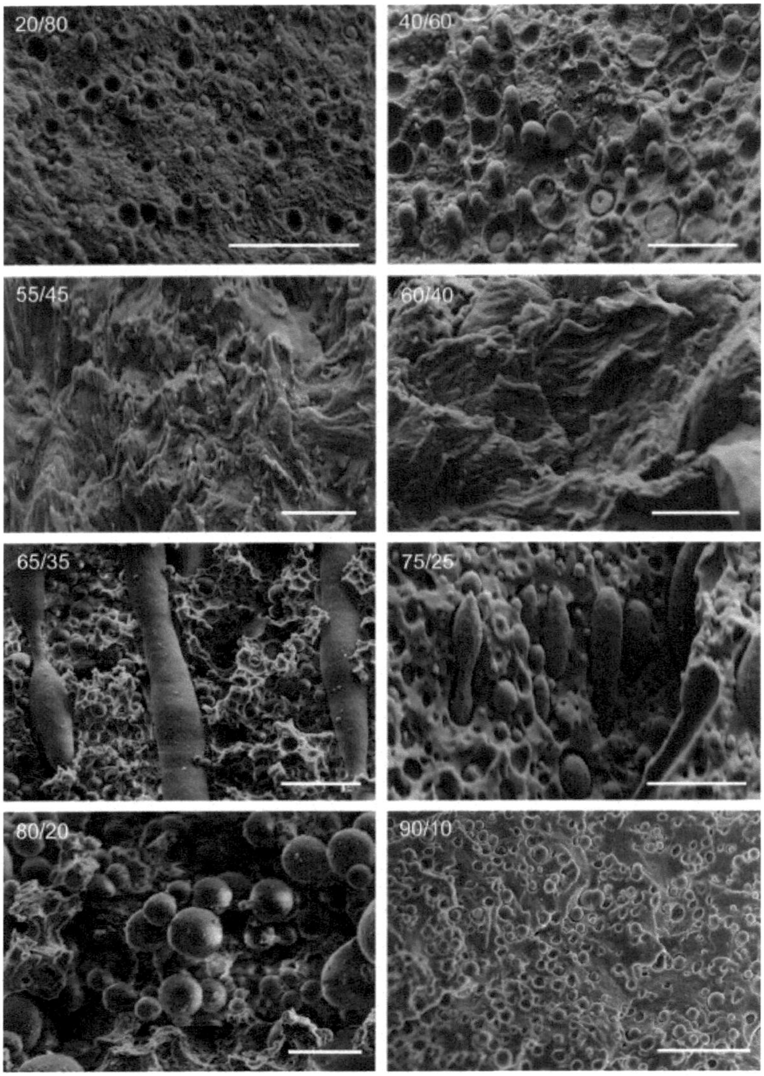

Figure 5.3 Scanning electron microscopy images of the PLA/PA11 blends (scale bar = 5 µm). Reproduced with permission from [9]. Copyright 2011, Elsevier Science Ltd

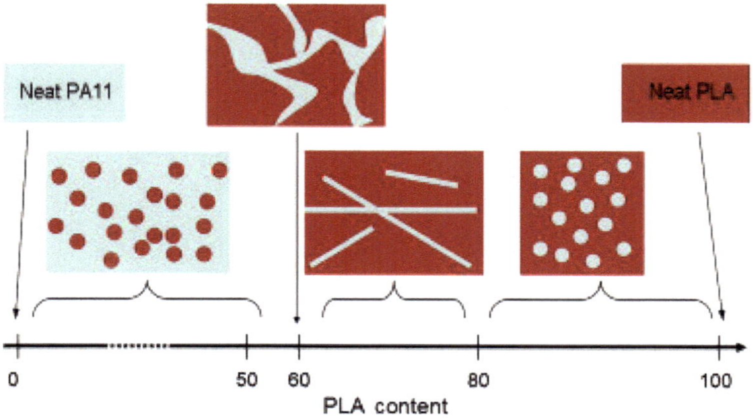

Figure 5.4 Schematic of the morphology evolution of the PLA/PA11 blends as a function of composition. Reproduced with permission from [9]. Copyright 2011, Elsevier Science Ltd

PA11 manifests as a spherical dispersion inside a PLA matrix when PLA percentages <80%. Once again, the dimension of the PA11 droplets spans from µm to sub-µm [9]. Their research indicated that excellent inherent compatibility of binary PLA/PA11 immiscible blends might be formed exclusive of the need for a compatibilizing agent, indicating that good self-compatible binary immiscible blends can be produced by simply altering the component composition.

So, it can be seen that appropriate processing using melt compounding can be utilized to prevent or allow phase separation and recrystallization in melt compounded polymeric blends.

5.2.2 Solution Blending

In solution blending of polymer blends, as illustrated in Figure 5.5, the polymeric matrices are dissolved in a suitable solvent while being vigorously stirred in this process. Finally, the mixture is recovered using the evaporation of the solvent or a nonsolvent. The fundamental disadvantage of this technology is that it is only appropriate for polymer matrices that are solvent-soluble. The recovery of the solvent is also difficult for the widespread application of this solution mixing method [1, 4].

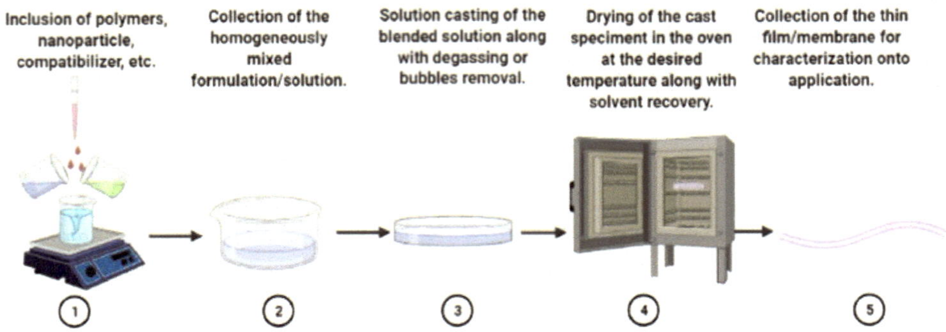

Figure 5.5 Schematic illustration of solution blending onto film casting of polymeric blended films/membranes (created with BioRender)

It thus is among the most basic strategies available, and it is quite popular among users. The polymer mixture is cast from a suitable solvent in this procedure, which includes the following stages:

(i) For mixing, the preferred component polymeric matrices are chosen.

(ii) A specific solvent is used to dissolve the specified polymers. It should be mentioned that the selection of solvent is critical and performs a significant part in the solution casting procedure.

(iii) Stir the polymer dope solution mixture for a specific length of time to obtain a homogeneous solution.

(iv) If necessary, the inclusion of binders as well as compatibilizers is carried out.

(v) At the end of the process, the final product is collected and described.

By using solution blending, immiscible polymer blends can be compatibilized along with reduced phase separation through the use of organic copolymers (block, random, or grafted) [10], reactive polymers, ionic or co-solvent interaction, the inclusion of nanoparticles, etc. [10]. The shape of solution cast combinations of telechelic polymers end-capped with tertiary amines as well as telechelic terminated polymers through one or the other carboxylate moieties or sulfonate closely mimics that of block copolymers generated via covalent bonding of binary different macromolecular chains [10].

This approach has been explored by Jang et al. [11] in their study of composition dependency of phase instability as well as cloud point in polystyrene-polyethylene solution-blended mixtures (cyclohexyl methacrylate). With the aid of a solution mixing approach, immiscible blends having high phase separation can be made miscible. For instance, some polymeric materials that are shown to be miscible with PS, for example, include ether-containing polymeric matrices, such as poly(1,4-dimethyl-*p*-phenylene oxide) (PPO) as well as poly(vinyl methyl ether) (PVME). PS blends are typically immiscible with ester-comprising (carbonyl functionalities) polymeric matrices or acrylic-based polymers. PS/PMMA blends are reported to be immiscible as well as incompatible. Shultz and Young stated that a single-T_g homogeneous PMMA/PS blends may be created by rapidly freezing their blend solution in a solution of naphthalene followed by room temperature naphthalene sublimation [12]. Research shows that the phase consistency of PS blends can be unstable or kinetically reliant when solution blending/casting of polymeric blends is utilized [12].

Jin et al. [13] explored self-construct thermal conductive filler connectivity in boron nitride nanosheets (BNNSs)/epoxy/polyetherimide composite materials by reaction-triggered phase-separation (see Figure 5.6). During the solution blending and casting curing procedure, the homogenous polymer mixture will split into two immiscible polymeric phases, known as reaction-induced phase separation (RIPS) [13]. The phase development of the dynamic asymmetric mixture DGEBA/PEI (100:30) (N-M00), as reported in their paper (Figure 5.6(a–d)), was observed by an optical microscope coupled temperature spike around 150 °C. A uniform phase was detected at the initial phase-separation (Figure 5.6(a)), confirming that the precursor DGEBA is combined with PEI as a suitable solvent. At 600 s, the phase-separation begins as a micro-dimensioned co-continuous phase architecture (Figure 5.6(b)). Figure 5.6(c,d) depict a phase inversion architecture (PEI-rich phase as matrix), wherein the lighter region reflects the epoxy-rich phase and the darkened region reflects the PEI-rich phase, as validated using SEM images. After 1590 seconds, the micro-sized phase inversion architecture was stabilized as the architecture of the later phases. The impacts of varied BNNS percentages on RIPS were also investigated further, and indeed the final phase morphologies of each hybrid are depicted in Figure 5.7. In comparison to the non-BNNSs-reinforced polymeric mix (Figure 5.7(a)), the addition of BNNSs-NH_2 speeds the grain growth of the epoxy-rich component, the microphase inversion architecture becomes more prominent, as well as the periodic distance between the dispersed epoxy-dominated phases, which grew as the BNNSs-NH_2 concentration increases (Figure 5.7(b–f)). The BNNSs-NH_2 linear acceleration effect is due to the robust interaction amongst mono-bond NH_2 functionalities on the BNNSs surface as well as the DGEBA precursor, which acts as a crosslinking throttle comparable to chemical cross-linking positions, changing not only the morphological characteristics but also the dynamics of binary mixture phase-separation [13].

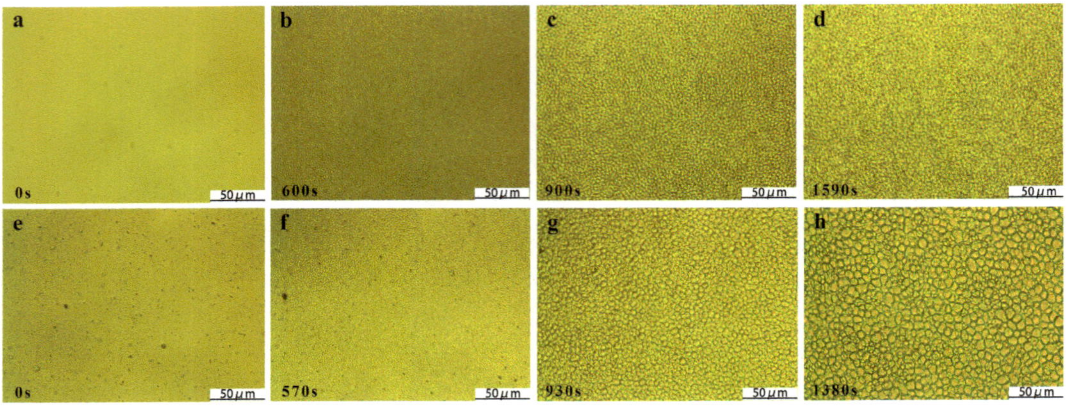

Figure 5.6 Morphology evolution of DGEBA/PEI30 (a–d) and 0.5 wt% BNNSs-NH_2/DGEBA/PEI30 (e–h). Reproduced with permission from [13]. Copyright 2020, Elsevier Science Ltd

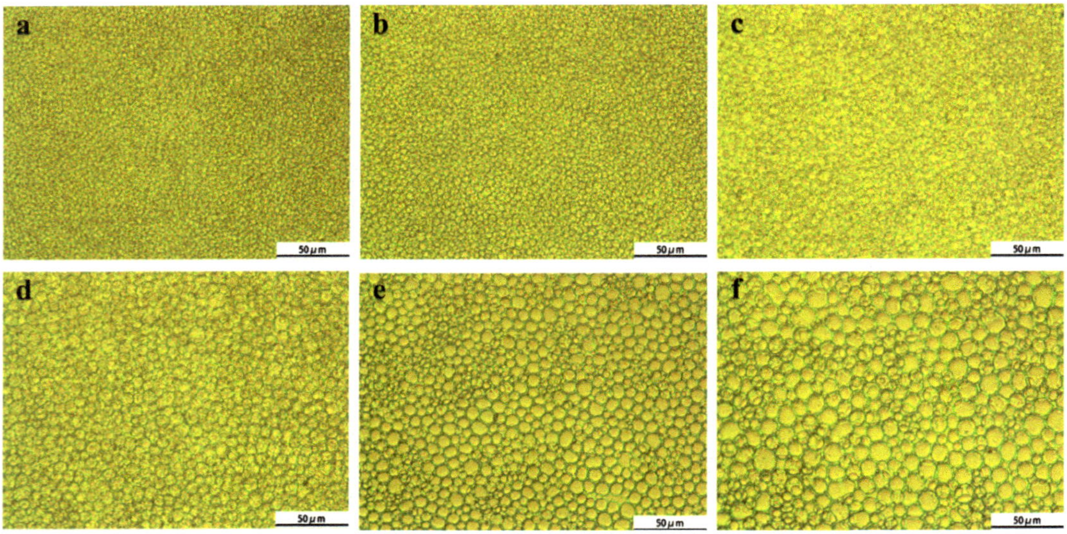

Figure 5.7 Optical micrographs of final morphology of DGEBA/PEI30 (a), and DGEBA/PEI30 with 0.1, 0.3, 0.5, 0.7, 1.0 wt.% BNNSs-NH_2 (b–f). Reproduced with permission from [13]. Copyright 2020, Elsevier Science Ltd

This kind of phase growth via RIPS is mostly more effectively achieved by the adoption of the solution blending/mixing approach. RIPS approach is vital towards the fabrication of systems for specified applications like enhanced thermal conductivity, electrical conductivity, dielectric properties, etc. [13].

5.2.3 Latex Mixing

This technique is majorly used for the preparation of elastomer blends for the purpose of enhancing the property performance of the final product. With regard to latex blending, the adopted polymeric elastomers are blended/mixed in their latex form [14, 15]. Elastomer polymeric miscible blends are fabricated via various diverse approaches including dry, solution, as well as latex blending, leading to extraordinarily stable fine dispersions. Latex interpenetrating polymeric networks are a form of polymeric matrix blend that is created by swelling crosslinked grain latex molecules of polymer 1 with monomer 2, plus crosslinking agents, and afterwards polymerizing monomer 2 in situ. Polymers 1 and 2 are normally incompatible to varying degrees, similar to polymer blends in a broad sense, as well as phase separation. Corish and Powell [16] have reviewed latex blending, like latex blending of natural rubber/butadiene rubber (NR/BR). Because of the low viscosity, solution mixing produces coarser dispersions because coalescence promotes rapid thermodynamic phase expansion. The quick phase growth could be attributable to the low viscosity of the solution as well as the lack of surfactants on the periphery of the polymer latex particles. This was observed by Walters and Keyte in natural rubber/styrene-butadiene rubber (NR/SBR) mixtures [17]. When the two rubbers were solution blended, they discovered a mixture of SBR encapsulated with NR. Even quick precipitation with a nonsolvent resulted in macro heterogeneity of the same type. Because the polymer molecules are produced in situ, there is a chance for blending at the molecular level; blending during solution polymerization is said to give better dispersion than solution blending. This aids the assimilation of carbon black as well. Solution and latex blending hybrids are also employed. The elastomer blend is retained in solution in one variant, while the black is disseminated in water in another. The black creates a master batch by transferring it into the elastomer. The black dissolved in a solvent is blended with latex in the reverse procedure. Dry blending in an intense mixer such as a mill, Banbury, or extruder produces the majority of elastomer mixes. This facilitates the mixing of compounding materials in a single process. Mastication facilitates further blending by reducing viscosity and allowing for reaction (block or graft polymerization) and covulcanization. According to Evans and Patridge, combining all components in one stage is preferable [18, 19]. Chemical aided blending, in which reactive monomers are introduced to one or both components, results in superior blends since the compatibilizer is formed in situ. Such graft polymer synthesis, according to Angier and Watson, happens in masticated mixes of NR, chloroprene rubber (CR), nitrile-butadiene rubber (NBR), SBR, and BR [20, 21].

Through latex compounding of immiscible phase-separated polymeric blends, poor integration among polysiloxane/acrylic resins in polymer mixtures [22] has been circumvented using a range of methodologies, including grafting of silanes on top of acrylic particles to reduce the interfacial tension amongst two diverse phases as well as cross-linking of the polymeric molecules during the coalescence phase of the blend formation process. Again, using latex compounded immiscible polymer blending technique, high strength, film-forming, transparent PMMA/PBA phase-separated coalesced latex blend has been fabricated by Lepizzera et al. [23]. They postulated that the contacts amongst the hard latex molecules control the ability of the blend to form films [23]. This shows that engineered phase-separated coalesced latex blend particles could be utilized for the preparation of crack-free transparent films for advanced applications (Table 5.1).

Table 5.1 Latex-Based Phase-Separated Blends and Applications

Latex blend	Phase separation controlling factor	Structure if available	Property enhanced	Proposed application	Ref.
PMMA/PBA	Hard latex grain size	-	Enhanced film formation	Biphasic thin films	[23]
Acrylic/PDMS	wt.% of PDMS	Coalescence	Enhanced film formation	Dirt resistance	[24]
Polyurethane/ PMMA	Gel-sol criterion is a viable route to PSAs with controlled adhesive properties	-	Enhanced adhesion	Pressure-sensitive adhesives (PSAs)	[25]
PBA-co-MMA	Polymer molar mass	-	Latex paints of low VOC composition	Latex coatings	[26]
Chitosan/natural rubber	CS ratio variation	Laminar	Enriched mechanical properties	-	[27]
Poly(butyl methacrylate-co-butyl acrylate-co-ethylene glycol dimethacrylate)	Cross-linking	Laminar	Polymer chain diffusion	Latex coatings	[28]
Poly(vinyl acetate–butyl acrylate)	Blend ratio	Co-continuous	Polymer chain diffusion	Thin film coatings	[29]
n-Butyl acrylate–methyl methacrylate	*n*-butyl acrylate (BA) volume fraction (copolymerization process as well as the monomer composition)	Drops	-	Controlled copolymer polymerization process	[30]
PMMA/PS	Heating rate (thermal annealing)	Ordered microphases, fibers, co-continuous, double emulsion, drops, etc.	Morphology stabilization (improved blend miscibility)	-	[31]

Latex blend	Phase separation controlling factor	Structure if available	Property enhanced	Proposed application	Ref.
Starch/natural rubber	NR wt.% and glycerol content	Co-continuous	Improved flexibility	-	[32]
PMMA-P(BMA-co-BA)	The two particles size ratio, surfactant removal from the systems, as well as film annealing after drying	Drops/spheres and in some blends co-continuous	Enhanced transparency	-	[33]
PB/PS and poly(ethyl methacrylate)/poly(*n*-butyl acrylate)	Incompatible blend ratio as a function of temperature	Core-shell/double emulsion	Film formability	-	[34]

Considering available works of literature, it is seen that majority of latex compounded polymeric blends favor laminar or co-continuous morphologies arising from the phase separation within the blend architecture. However, with the help of controlled process parameters during latex compounding, it is possible to engineer phase separation-derived morphologies of choice. Therefore, one can carefully select the appropriate processing method per the targeted structural architecture towards preferred applications.

5.2.4 Interpenetrating Polymer Networks (IPNs)

Interpenetrating polymer networked mixable blends have been reportedly explored by diverse researchers for the preparation of polymer blends [21, 35–37]. IPNs are a combination of ≥2 polymers in networks where a partial interlacing on the molecular scale is present in the matrix, but the polymers are not bonded to each other. They are synthesized via diverse approaches, simultaneously or sequentially. This approach is similar to the mechanochemical approach, except it does not rely on external mechanical force. The polymer mix system contains ≥2 polymers deeply embedded in each other, and to break up the IPN requires the breaking of chemical bonds which demands a lot of energy [38]. The components in these kinds of blends are produced in such a manner that the polymer chains are intertwined, although not by chemical bonds. As a result, unique procedures are often necessary to create such polymeric systems, while simple mixing would not suffice to build an interpenetrating polymer network (IPN). One or more polymer networks, including linear or nonlinear branch polymers, make up the semi-interpenetrating polymer network (SIPN). This system is defined by the molecular implanting of no less than one network of the polymeric matrix into the other polymer. The polymeric part of the network can be detached from the network in this situation without disrupting any chemical bonds [38]. Pseudo-interpenetrating polymer networks (PIPN) are created when a polymer is partially penetrated into the networked blend system. As a result, the polymer, as well as network components, are distinct, their properties are preserved to a larger extent, as well as separation is simpler than with SIPN [38].

Hydrogels specifically may be categorized into homopolymers, copolymers, semi-interpenetrating networks, as well as interpenetrating networks based on the polymeric material. Homopolymer nanogels are made up of the same repeating units and have a cross-linked as well as uncross-linked architecture depending on the type of monomer as well as the polymerization approach. Copolymeric hydrogel is made up of two or more distinct monomer units; each one with at least one hydrophilic constituent is an example in this regard. It has been shown that a copolymer-based immiscible phase-separated hydrogel made of PLGA-PEG-PLGA can dissolve the hydrophobic drug indomethacin because when the drug is loaded within the copolymer micelles, its solubility increases [39]. The polymerized polymeric structure, in turn, results in a SIPN which results in sustained release of the loaded drug [39].

SIPN polymeric biomaterials are fashioned mostly by non-chemical infiltration of a given linear polymeric chain into yet another cross-linked polymer structure. These structures are mostly applied for advanced materials engineering in the biomedical field. Figure 5.8 depicts the preparation of a silk fibroin@PA semi-interpenetrating networked hydrogel, used as a template for drug release [40]. The SEM examination of fibroin/PAAm SIPN hydrogels confirmed uniform pore structure, Figure 5.9. The microstructures showed that polyacrylamide, as well as fibroin, produce an uninterrupted phase in entire-polymer mixtures, with no visible phase separation. All across the formulations, relatively uniform pore distribution was evidenced. These same pores are open-ended, well-connected, and have well-defined interfaces. A 70/30 semi-IPN demonstrates denser interfacial partitions. The mean size of the SF/PAAm semi-IPN of 70/30 formation pore was 100 to 14 μm. Pore size had also been reduced to 50 to 11 μm, 30 to 10 μm, but also 15 to 6 μm in the case of SF/PAAm mixtures with constituents of 60/40, 50/50, as well as 0/100. A significantly greater polyacrylamide content led to an extremely high pore density and with a higher number of pores per unit area [40]. It is possible that as crosslinking depth increases, water content is reduced, developing into lesser ice crystal materials that are then expressed by smaller pores upon lyophilization and vice versa.

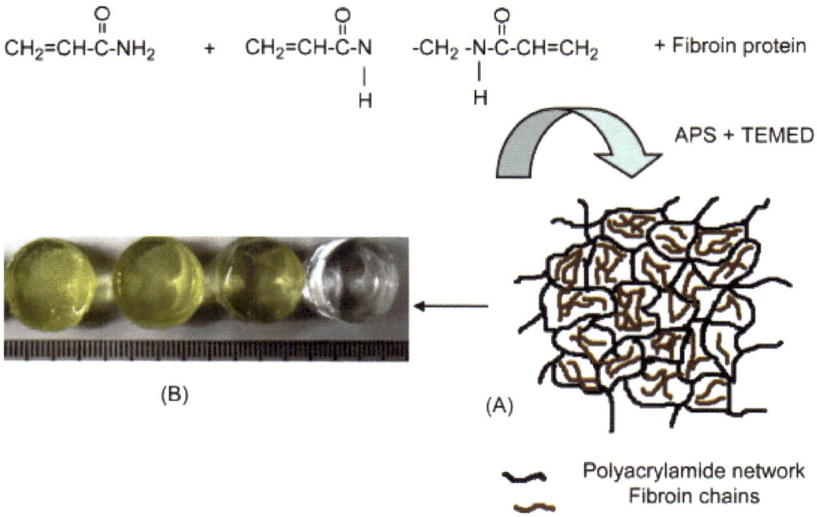

Figure 5.8 Semi-interpenetrating network made up of polyacrylamide polymer and silk fibroin chains Incorporation of fibroin improves the elasticity and softness of the hydrogel and helps in sustained release of drug (A) Schematic representation of the formation of SF/PAAm semi-IPN network hydrogels (B) Typical appearance of different SF/PAAm semi-IPN hydrogels after removing from molds. Reproduced with permission from [40]. Copyright 2009, Elsevier Science Ltd

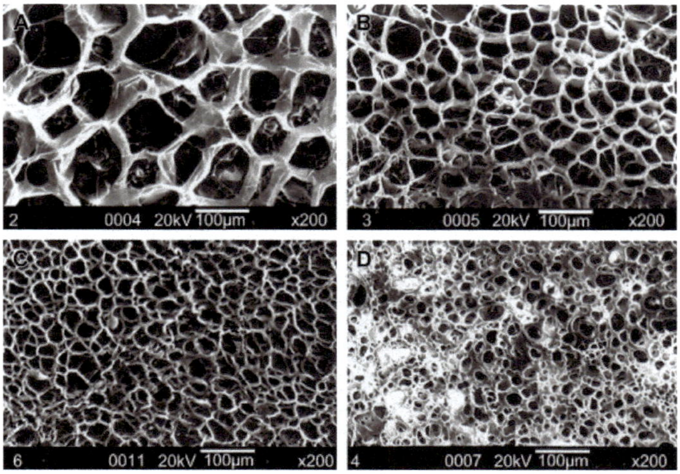

Figure 5.9 SEM photographs showing cross-sectional morphology and pore size within semi-IPNs with composition in the ratios of (SF/PAAm): (A) 70/30, (B) 60/40, (C) 50/50, and (D) 0/100. Reproduced with permission from [40]. Copyright 2009, Elsevier Science Ltd

In recent decades, researchers have become more interested in IPNs due to their mechanical characteristics, particularly the toughness of interpenetrating hydrogel. This same toughness of hydrogels is reportedly determined by the combination of covalently as well as ionically crosslinked polymeric matrix networks [41]. Lin et al. [42], on the other hand, created a dual interpenetrating network hydrogel without the formation of any covalent bonds using cellulose and polyacrylamide polymers. The mechanical strength of these IPN structures integrated with hydrogel is substantially enhanced. According to Lee et al. [43], PEG-diacrylate, along with a modified gelatin IPN system loaded with gold porphyrin (AuP) hindered tumor growth as well as reduced angiogenesis in modelled mice carrying human lung cancer xenografts.

5.2.5 Graft Copolymerization

The utilization of graft copolymerization for the preparation of phase-separated polymer blends has been reported, explored, and discussed as per literature [44–47]. One of the strategies adopted for the compatibilization of polymeric blends, especially phase-separated systems, is the use of graft copolymerization [48]. To this end, a group of researchers adopted graft copolymers that improve the strength of polymer composites designed using a combination of computer modelling and experimental analysis where they reported that the copolymers synthesized possessed side chains that form a "molecular velcro" across the boundary between phase-separated regions, effectively binding the regions together [44]. This behavior increased the structural integrity and mechanical qualities of the material dramatically even as their process considerably expanded the choice of copolymers that may be utilized to build high-strength polymer blends since the side chains can be made from a wide range of substances [44]. Deuterated poly(ethyl acrylate) (*d*-PEA) plus deuterated PMMA (*d*-PMMA) were used as the immiscible A and B homopolymers for the studies (*d*-PMMA): PEA@PS and PMMA@PS graft copolymers with a polydisperse PMMA as well as PEA backbone along with an average of single monodisperse, randomly placed PS prong on each backbone were produced, but only the homopolymers were deuterated for the experiment [44]. This approach has also been utilized by other researchers like Zhang et al.[45] for the fabrication of immiscible polymer blends where phase-separated PMMA@styrene acrylonitrile (SAN) copolymer (60/40) blend with the aid of $nSiO_2$ grafted with PMMA brushes are explored. They revealed in their study that the PMMA-rich phase contained NPs with brushes that were significantly bigger than the matrix PMMA chains making them easily distributed throughout the PMMA as well as SAN-dominated areas, which they verified to be controlled by the entropy-controlled interfacial energy amongst the PMMA molecules as well as chains of the brush [45].

The grafting of polymeric molecules onto nanoparticles, such as MWCNTs [49] and nanoclays [50], has been exploited by researchers with the aim of enhancing the compatibilization of immiscible phase-separated polymeric blends. Zhang et al.

[49] synthesized a MWCNTs-g-PMMA via Cu(0)-mediated living radical polymerization (LRP) as depicted in Figure 5.10. The predicted wetting coefficient "ω_a" as well as location of the MWCNTs-g-PMMA in their reported blends agreed well with experimental results.

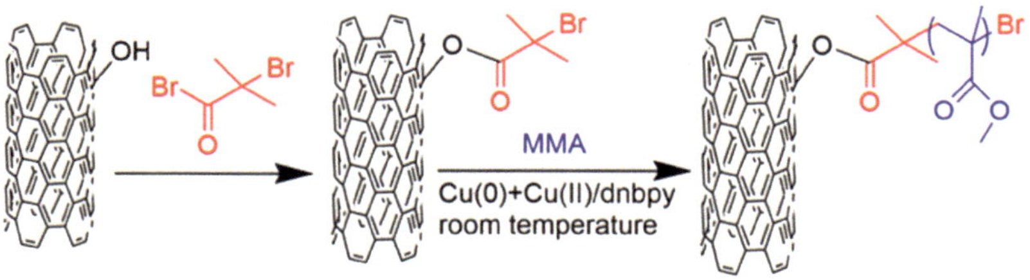

Figure 5.10 Schematic for the synthesis of MWNTs-g-PMMA by Cu(0)-mediated LRP. Reproduced with permission from [49]. Copyright 2022, Elsevier Science Ltd

The MWCNTs-g-PMMA were found to be primarily in what seems like a thermodynamic equilibrium condition just at the boundary of several polymeric blends, including PMMA@SAN, PMMA@PS, PMMA@PVAc, as well as PMMA@MSAN blends, as reported in Figure 5.11, demonstrating that the placement of polymer-brush-grafted nanoparticles might well be unaffected by chemically diverse matrix types. As a result, the authors not only prove at the mechanistic level the contribution of entropy to the selective location of polymer-grafted nanoparticles in the blended matrix, but also provide a route as a reference template to formulate innumerable composite materials having nanoparticles positioned at the interface for desired applicability [49].

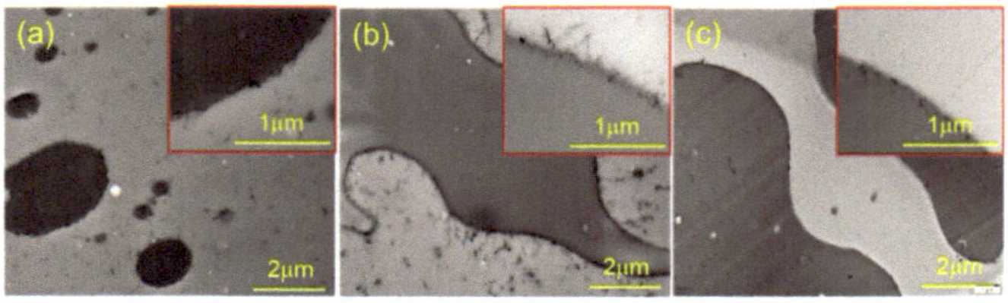

Figure 5.11 Transmission electron microscope images of 1.5 wt.% MWNTs-g-PMMA filled (a) PMMA/PS (50/50), (b) PMMA/PVAc (50/50), and (c) PMMA/MSAN (50/50) blends annealed for 1.5 h at 200 °C. The PMMA phases appear as bright regions, while the dark regions refer to the PS, PVAc, and MSAN regions, respectively. Reproduced with permission from [49]. Copyright 2022, Elsevier Science Ltd

The adoption of this approach for the preparation of blends along with the use of innovative novel supplementary techniques, will surely lead researchers towards the manufacture of unique blends with excellent performance for end-user applications.

5.2.6 In Situ Polymerization

The preparation of blends via the technique of in situ polymerization is amongst the most utilized approaches in this area [51–55]. The preparation of PMMA/PVAc blends via in situ polymerizing MMA in the presence of PVAc, with compositions of 95/5, 90/10, 80/20, 70/30, and 60/40 (wt./wt.) has been explored [52], with consideration to MMA weight, 0.1 wt.% benzoyl peroxide was utilized as the initiator: the procedure of multistage polymerization was applied. To avoid the Trommsdorff–Norrish phenomenon and achieve a critical viscosity level, the pre-polymerization of MMA was first carried out at 80 °C with continuous stirring for 30 min; afterwards, the solution was cooled and poured into a mold, which consisted of two polished glass plates divided by Teflon pieces at the four corners and sealed with adhesive tape [52]. To complete the polymerization reaction, the mold was placed in an oven at 40 °C for 20 hours, 80 °C for 4 hours, and 100 °C for 4 hours. In comparison to ordinary polystyrene, the PMMA produced had a M_n of 685,700 as well as weight average molecular weight (M_w) of 944,500 g/mol, as assessed by gel permeation chromatography (GPC) [52]. The 80/20 in situ polymerized blend was dissolved in MMA and other solvents, including acetone, THF, chloroform, cyclohexane, dioxane, and *N,N*-dimethylformamide (DMF), and then cast to make the film samples to study the reliance of phase behavior on solvents. The bulk of the solvents was evaporated at ambient temperature (for acetone, THF, and chloroform) or at 40 °C (for MMA, DMF, cyclohexanone, and dioxane); the remaining solvents were removed in vacuo for two weeks [52]. In another work, Zheng et al. [54] have shown that polymerization in situ, that is, melt polymerization between the diglycidyl ether of bisphenol A (DGEBA) with bisphenol A in the presence of poly(acrylonitrile-co-styrene) (SAN), can be used to make poly(hydroxy ether of bisphenol A)-based polymeric mixtures incorporating SAN. Starting with a homogenous ternary combination of SAN@DGEBA@bisphenol A, the polymerization reaction produced phenoxy@SAN blends with SAN content up to 20 wt.%. The solubility behavior of the reactive blend system and FTIR measurements both showed no intercomponent reaction occurrence. To evaluate the phase structure of the as-polymerized blends, researchers used DSC, DMA, as well as SEM. All the polymeric mixtures have distinct glass transition temperatures (T_gs), indicating that they were phase-separated. Also, the blends had excellently distributed phase-separated structures, according to the morphological analysis. Very small SAN particulate spheres (Ø = 1–3 mm) were observed to uniformly disseminate into a continuous phenoxy polymeric matrix for the mixtures containing SAN content <15 wt.%. Also, the exceptional structure was generated via phase-separation-driven polymerization [54]. Mechanical testing revealed the mixtures containing 5–15 wt.% SAN had significantly improved tensile characteristics and Izod impact strength compared to the composites manufactured using traditional procedures [54]. Zheng et al. [55] in another report, reported that in situ polymerization, i.e., melt polymerization of DGEBA and bisphenol A in the presence of PVAc, was used to create blends of poly(hydroxyether of bisphenol A) (phenoxy) and poly(vinyl acetate) (PVAc). The polymerization reaction began with an evenly mixed ternary mixture of PVAc@DGEBA@bisphenol A; as the polymerization progressed, phase separation was triggered by the reacting mixture. The solubility, FTIR, DSC, DMA, as well as SEM were used to analyze the phenoxy@PVAc mixtures having PVAc content towards 20 wt.%. During the in situ polymerization, no intercomponent reaction occurred [55]. The phase-separated extremely fine structure was generated by the polymerization mixing approach; all of the blends had separate T_gs. Mechanical testing revealed that the produced blends improved their mechanical properties significantly, particularly for impact strength, which was ascribed to the creation of the fine phase-separation architecture for the duration of the polymerization. Due to the synergistic contribution of the two phases in energy transfer, the thermal stability of the PVAc-dominant areas within the mixtures was improved when compared with the virgin PVAc, according to the TGA of the blends [55]. Ogawa et al. [56] demonstrated that poly(silmethylene)-based polymeric blends were produced by polymerization in situ, 1,1,3,3-tetraphenyl-1,3-disilacyclobutane in flexible organosilicon polymers with or without a catalyst. A different approach, in which blends are made by synthesizing a flexible polymeric matrix in the presence of a rigid one proved equally encouraging. DSC, dynamic mechanical analysis, and solvent extraction were used to characterize the resultant polymer blends. When in situ copper-catalyzed polymerization was used, no chemical interaction between component polymers of blends formed by in situ bulk polymerization, but synthesis of block or graft copolymers including poly(diphenylsilmethylene) as well as flexible polymers was indicated. Microscopic inspection revealed a morphological variation between materials created using the various procedures [56].

In situ polymerization is reportedly broadly utilized for the fabrication of phase-separated polymeric blends, especially in latex compounding [22–25, 29]. Though phase separation in these systems is mainly process controlled especially via the control of blend ratio, processing temperature, or solvent polymer interaction.

5.2.7 Mill Mixing

A facile and straightforward method of blending is that the mixing components are blended by milling as well as grinding. For this purpose, various milling tools as well grinders, are used. The precursor materials are ground to produce finely divided powders, which are then combined to attain homogeneous blending at the microlevel. Additional operations are then performed on the product to obtain the preferred polymeric blend items [57]. The Banbury mixer, sometimes referred to as even Master Mixer, is commonly utilized to mechanically blend polymer components and as such, is appropriate for polymer blend manufacture [57].

A general way of preparing rubber-based polymeric mixtures is mill mixing. Mixing can take place amongst two parallel arranged rotational hollow metallic pistons. The nip expanse amongst two rollers could be tuned by adjusting the roller gaps according to the compound uniformity. Several researchers have used this approach for the preparation of immiscible polymeric blends for advanced application within the present century [58]. Researchers, as well as industrialists, have adopted the mill mixing technique for the preparation of diverse polymeric blends in recent decades [59, 60]. For instance, Bazli et al. [59] investigated silicone rubber (SR) and ethylene-propylene-diene monomer (EPDM) blends obtained through two-roll mill mixing with different weight contents of nanoclay (Cloisite 15A) in the SR, EPDM, and SR/EPDM hybrid nanocomposites for outdoor polymeric insulators. Furthermore, the breakdown voltage distribution was fitted to the Weibull distribution, and the scale and shape parameters were estimated. The polar groups of the clay particles improved the polarization capability of the nanocomposites.

Furthermore, dielectric breakdown strength results revealed an increase in dielectric strength proportional to clay content. Finally, while all nanocomposites surface and volume resistance decrease, the electrical resistance remains very high [59]. The mill mixing process was shown in this work to effectively enhance the uniform distribution of the reinforcing fillers within the blend, as depicted in Figure 5.12. It was postulated that the link between the phases elasticity coupled with the viscosity possibly explains the transition from spherical EPDM drops to oval shapes. Due to the obvious nanoclay existence there in dispersed phase plus matrix, as well as the resulting drop in viscosity, the structure advances forward into a co-continuous structure. Furthermore, the entrapment of nanoclay within EPDM caused a spike in the elasticity of this phase, a decrease in shape relaxation correlated well with droplets, as well as the retention of the egg-shape globules due to their high contacts. As a result, nanoclay demonstrated a possible compatibilizing impact in such nanostructures [59].

The mixing operation is the most ancient of all processing operations. Thomas Hancock received the first patent for an internal mixer in 1823 [61]. In 1876, Freyburger patented a more effective, counter-rotary twin-rod internal blender, which was modified by P. Pfleiderer [62] and B. F. Banbury [62]. E. M. Chaffe [60] invented the two-roll mill, which was manufactured by Farrel Foundry and Machine Co.

Mill mixing is a more utilized approach for the preparation of elastomeric polymer-containing blends [63, 64]; this technique enables the homogeneous blending of the polymer matrices along with the diverse additives leading to effective structural architecture as well as reduced phase separation for better compatibility and excellent final product properties.

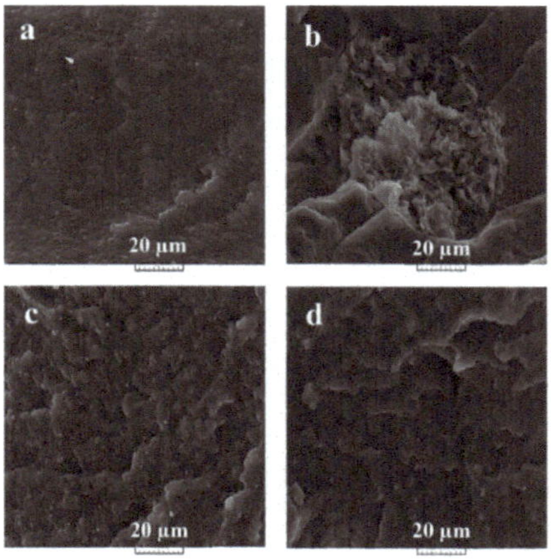

Figure 5.12
SEM images of the samples, (a) E100C9, (b) Q100C9, (c) Q70E30C0, and (d) Q70E30C9. Reproduced with permission from Bazli et al. [59]. JCC Research Group (Location: Iran) 2021

■ 5.3 Characterization of Phase-Separated Polymer Blends

Different characterization procedures utilized for phase-separated polymer blends are given in Table 5.2 underneath; a detailed chapter is presented in **Chapter 6**.

Table 5.2 Approaches to the Characterization of Phase-Separated Polymer Blends

Characterization methods	Utilization	Ref.
Atomic force microscopy (AFM)	Morphology, microstructure, phase separation, surface coarseness, etc.	[65]
Differential scanning calorimetry (DSC)	Amongst the most extensively utilized thermal analytical characterization approaches for polymers as well as polymeric blends is DSC. Herein, the deviation in the quantity of heat needed to raise the ample temperature as well as the reference, is computed as a function of heat flux/temperature. The reference temperature, as well as the sample, should grow in a linear manner with respect to time ramped at a certain rate during the experiment. It is worth noting that the reference sample should be scanned across a wide temperature range to determine its heat capacity. In terms of heat flux vs time or temperature, DSC provides experimental results in the form of a DSC graph. This method is the most effective for determining the T_g of the system as well as enthalpies (H). Furthermore, the DSC-identified depression within the obtained melting temperature could be utilized to ascertain the miscibility of the blend system. Similarly, the DSC graph can be used to determine the polymer/polymer interaction values.	[45, 52, 54–56, 66–68]

Table 5.2 Approaches to the Characterization of Phase-Separated Polymer Blends *(continued)*

Characterization methods	Utilization	Ref.
Dynamic mechanical analysis (DMA)	DMA could be utilized to investigate the viscoelastic character of polymers, which are of great interest to researchers. In DMA, sinusoidal stress is employed in polymeric blends. Stress causes some changes in polymer blend, which are then used to calculate strain and, ultimately, complex modulus. DMA could also be adapted to determine the T_g of a specified polymer/blend. Aside from that, thermomechanical analysis (TMA) is useful for identifying transitions that are akin to specific molecule motions.	[67, 69]
Dielectric relaxation spectroscopy (DRS)	This approach studies the polymer miscibility and crystallization kinetics of amorphous and crystalline polymeric blends	[67, 70]
Ellipsometry	Assessment of the interfacial cross-sectional area of two polymers, as well as the T_g (of thin films)	[71]
Fluorescence microscopy	This approach investigates polymer blend system structural analysis	[72, 73]
Positron annihilation spectroscopy (PAS)	The miscibility, as well as free-volume distributions of polymeric blends, is studied by this technique	[74, 75]
Electron paramagnetic resonance (EPR) spectroscopy along with forward recoil spectrometry (FRES)	The analysis of surface, as well as interfacial behavior along with polymer blend processes, is achieved by this approach	[76]
Raman imaging	Assessment of the phase syllable structure is carried out via this process	[76]
Solid-state nuclear magnetic resonance (NMR) spectroscopy	Assessment of intermolecular as well as intramolecular interactions, miscibility, as well as microstructures of polymeric blends can be studied using this approach	[76, 77]
X-Ray diffraction (XRD)	Crystallization studies and structural characterization (degree of crystallinity, size of crystals). This is among the most feasible crystallinity characterization approach for polymer blends. The organized atoms in XRD induce the incident X-ray beam to be diverted in a variety of paths, wherein the angles, as well as intensities, are obtained in a graph known as a crystallograph. The said crystallograph is packed with crystal lattice as well as unit cell details. The electron density is used to find the average locations of the atoms within the said crystal, along with their chemical bonds, disorder, as well as other data.	[68, 76, 77]

Characterization methods	Utilization	Ref.
Secondary ion mass spectrometry (SIMS) techniques, time-of-flight secondary ion mass spectrometry (ToFSIMS), and nanoscale secondary ion mass spectrometry (Nano SIMS)	External characterization of polymers (with particular emphasis on polymeric blends as well as interfaces), as well as molecular architectural details (like saturation, molecular weight distribution, branching, segmental length, functional groups, etc.)	[78–80]
Light scattering	Studies of crystallization, phase separation, cloud-point identification (to determine the phase periphery), and particle size assessment: this technique could be adopted to observe the phase separation phenomenon in polymer A/polymer B/solvent tertiary systems during solvent evaporation	[81]
Neutron scattering Ultrasound	Morphology, miscibility ultrasound polymer blend composition, flow extrusion instabilities measurement, injection molding process monitoring, polymeric molecule modification, immiscible blends compatibilization, nanoparticles dispersion within polymeric melts, morphological analysis, and so on.	[82]
Optical microscopy	Used to analyze the spherulite growth rate morphology, crystallization kinetics, etc.	[13, 73]
Mechanical test/ tensile tester	The most desirable aspects of polymer blends are their mechanical properties. Currently, great effort is made into creating mechanically strong products. The tensile tester can be used to test the mechanical strength of the products. Tensile testers, universal testing machines (UTMs), and the graphs that determine mechanical strength generally work on the stress vs strain theory, as well as young modulus could be attained from the charts which define mechanical strength.	[15]
Thermogravimetric analysis (TGA)	Utilized for the determination of thermal stability of polymeric blends	
Infrared (IR) spectroscopy	Constituent identification and evaluation of interfacial interfaces, as well as exploration of the interactions taking place between different polymeric matrices within the blend	[52 ,65, 68, 83]
Transmission electron microscopy (TEM)	TEM is an optical approach that transmits an electron beam via an ultra-thin specimen and also provides a wealth of details regarding the specimen. TEM micrograph(s) can be magnified as well as focused onto an attached imaging device to extract the data, and the assimilated data about the polymeric blend could be easily obtained. In most	

Table 5.2 Approaches to the Characterization of Phase-Separated Polymer Blends *(continued)*

Characterization methods	Utilization	Ref.
	cases, the TEM sample preparation approach is like the SEM sample preparation procedure. For TEM analysis, thin sections of polymeric blends are typically cut with a diamond knife supported by a UC6 microtome. Because polymeric blend specimens are comparatively poor conductors, the gold coating is typically applied.	
Rheometry	Utilized for the determination of flow properties, as well as the viscoelastic properties of polymer blends	[73]
Scanning electron microscopy (SEM)	SEM is a kind of microscopy in which sample images are created by scanning a sample surface with an electron beam. The electrons interact with the atoms of the sample generating a variety of signals that are sensed as well as converted into images by a micrograph computer. This micrograph comprises useful data about the surface topography of the sample along with its composition. This technique is commonly used to characterize the phase morphology of polymeric blends. In the case of polymeric blends, this characterization approach aids researchers in predicting component distribution. Furthermore, phase homogeneity can be predicted. The specimens are fractured cryogenically herewith by immersing them in liquid N_2. A brittle fracture is obtained while averting significant deformations on the surface under investigation.	[65, 67, 68]
Miscellaneous characterization techniques	Gradient HPLC approaches, like gradient polymer elution chromatography (GPEC), have been utilized in recent years to determine the compositional drift of copolymers or the examination of polymeric additives as a type of composition of polymeric mixtures. Molecular-weight-based separations, mechanism of adsorption, or types of precipitation can be done with the right columns and gradient conditions. Researchers can gather particular information about a material by using a customized detector. An evaporative light scattering detector (ELSD), for example, enables the use of a universal mass detector to execute solvent gradients and monitor both UV-absorbing as well as non-UV-absorbing polymeric specimens without disruptions to the baseline. The addition of a photodiode array detector (PDA) allows compositional analysis across the molecular weight distribution of many copolymers, which is useful for identifying components in polymer blends and quantitating polymer additives as well as other small molecules in conventional reversed-phase separations. The composition and structural details of the polymer blends are also determined using gradient gel permeation chromatography (GPC) [84]. Gomez et al. used a theoretical technique to investigate the thermodynamics of binary as well as ternary polymeric	[84, 85]

Characterization methods	Utilization	Ref.
	mixtures and compared their findings to investigational computations obtained using GPC [85]. The DSC-identified depression in the obtained T_m could also be utilized to ascertain blend system miscibility. Similarly, the polymer-polymer interfacial values could be determined using the DSC plot.	

5.4 Conclusions

This chapter is an inclusive, but not exhaustive, coverage of the preparation and characterization of phase-separated polymer blends. We have discussed selected examples of how the preparation of polymeric blends affects phase separation and morphological changes, and other properties of the polymeric blends or their composites. Vital systems adopted for analyzing the phase separation as well as miscibility of polymeric blends are DSC as well as DMA by means of assessing the blend T_g as a conventional approach to evaluating the miscibility of polymeric blends. A single T_g for is found for miscible blends, while phase-separated polymeric blends will possibly present two or more T_gs. DMA, as well as DSC, are the major approaches adopted for measuring the blends of individual components T_g.

References

[1] Moniruzzaman, M., Winey, K.I. Polymer nanocomposites containing carbon nanotubes. *Macromolecules* 39, 5194–5205 (2006)

[2] Huang, Y.Y., Terentjev, E.M. Dispersion of carbon nanotubes: mixing, sonication, stabilization, and composite properties. *Polymers* 4, 275–295 (2012)

[3] Tang, W., Santare, M.H., Advani, S.G. Melt processing and mechanical property characterization of multi-walled carbon nanotube/high density polyethylene (MWNT/HDPE) composite films. *Carbon* 41, 2779–2785 (2003)

[4] Khan, W., Sharma, R., Saini, P. Carbon nanotube-based polymer composites: synthesis, properties and applications. In *Carbon Nanotubes-Current Progress of their Polymer Composites*. Berber, M. (2016), IntechOpen, Chapter 1, DOI: 10.5772/61428

[5] Xie, X.-M., Sui, X. In *Compatibilization of Polymer Blends*. Ajitha A.R., Thomas S. (Eds) (2020), Elsevier, Amsterdam, 145–177

[6] Dekkers, M., Hobbs, S., Bruker, I., Watkins, V. Migration of polymer blend components during melt compounding. *Polym. Eng. Sci.* 30, 1628–1632 (1990)

[7] Mao, H.-j. et al. Constructing a microcapacitor network of carbon nanotubes in polymer blends via crystallization-induced phase separation toward high dielectric constant and low loss. *ACS Appl. Mater. Interfaces* 12, 26444–26454 (2020)

[8] Macosko, C.W. In *Macromolecular Symposia*. Vol. 194, 171–184 (Wiley Online Library, 2000)

[9] Stoclet, G., Seguela, R., Lefebvre, J.-M. Morphology, thermal behavior and mechanical properties of binary blends of compatible biosourced polymers: Polylactide/polyamide11. *Polymer* 52, 1417–1425 (2011)

[10] Koning, C., Van Duin, M., Pagnoulle, C., Jerome, R. Strategies for compatibilization of polymer blends. *Prog. Polym. Sci.* 23, 707–757 (1998)

[11] Jang, F.H., Woo, E.M. Composition dependence of phase instability and cloud point in solution-blended mixtures of polystyrene with poly(cyclohexyl methacrylate). *Polymer* 40, 2231–2237 (1999)

[12] Shultz, A.R., Young, A.L. DSC on freeze-dried poly(methyl methacrylate)-polystyrene blends. *Macromolecules* 13, 663–668 (1980)

[13] Jin, X., Li, W., Liu, Y., Gan, W. Self-constructing thermal conductive filler network via reaction-induced phase separation in BNNSs/epoxy/polyetherimide composites. *Composites, Part A* 130, 105727 (2020)

[14] Varamesh, A., Abdollahi, M., Khanli, H.H. Structure and properties of NR/BR blend/clay nanocomposites prepared by the latex method. *Polym. Sci., Ser. A* 55, 115–120 (2013)

[15] Abdollahi, M., Khanli, H.H., Aalaie, J., Yousefi, M.R. Structure and properties of natural rubber/butadiene rubber (NR/BR) blend/sodium-montmorillonite nanocomposites prepared via a combined latex/melt intercalation method. *Polym. Sci., Ser. A* 53, 1175–1181 (2011)

[16] Corish, P., Powell, B. Elastomer blends. *Rubber Chem. Technol.* 47, 481–510 (1974)

[17] Walters, M., Keyte, D. Heterogeneous structure in blends of rubber polymers. *Rubber Chem. Technol.* 38, 62–75 (1965)

[18] Mangaraj, D. Elastomer blends. *Rubber Chem. Technol.* 75, 365–427 (2002)

[19] Evans, L., Partridge, E. ACS Rubber Division., 83rd Meeting, 1963. *Rubber Age (N. Y.)* 94, 272 (1963)

[20] Alegria, A., Elizetxea, C., Cendoya, I., Colmenero, J. Anomalous dynamical homogeneity of the dielectric α-relaxation in miscible polymer blends of poly(epichlorohydrin) and poly(vinyl methyl ether). *Macromolecules* 28, 8819–8823 (1995)

[21] Silverstein, M.S. Interpenetrating polymer networks: So happy together? *Polymer* 207, 122929 (2020)

[22] Richard, J., Mignaud, C., Sartre, A. Stability and compatibility in blends of silicone and vinylacrylic polymer emulsions. *Polym. Int.* 31, 357–365 (1993)

[23] Lepizzera, S., Lhommeau, C., Dilger, G., Pith, T., Lambla, M. Film-forming ability and mechanical properties of coalesced latex blends. *J. Polym. Sci., Part B: Polym. Phys.* 35, 2093–2101 (1997)

[24] Khanjani, J., Pazokifard, S., Zohuriaan-Mehr, M.J. Improving dirt pickup resistance in waterborne coatings using latex blends of acrylic/PDMS polymers. *Prog. Org. Coat.* 102, 151–166 (2017)

[25] Mehravar, S., Ballard, N., Agirre, A., Tomovska, R., Asua, J.M. Relating polymer microstructure to adhesive performance in blends of hybrid polyurethane/acrylic latexes. *Eur. Polym. J.* 87, 300–307 (2017)

[26] Tomba, J.P., Ye, X., Li, F., Winnik, M.A., Lau, W. Polymer blend latex films: Miscibility and polymer diffusion studied by energy transfer. *Polymer* 49, 2055–2064 (2008)

[27] Magida, M.M., Ibrahim, S.M., Elnahas, H.H. Effect of gamma radiation on the characterization of chitosan/natural rubber latex polymer blends. *Int. J. Polym. Anal. Charact.* 25, 565–580 (2020)

[28] Tamai, T., Pinenq, P., Winnik, M.A. Effect of cross-linking on polymer diffusion in poly(butyl methacrylate-co-butyl acrylate) latex films. *Macromolecules* 32, 6102–6110 (1999)

[29] Oh, J.K., Yang, J., Rademacher, J., Farwaha, R., Winnik, M.A. Phase separation and polymer diffusion in poly(vinyl acetate–butyl acrylate) latex films prepared by batch miniemulsion polymerization. *Macromolecules* 37, 5752–5761 (2004)

[30] Erola, M.O.A., Suvanto, S., Pakkanen, T.T. Influence of compositions and polymerization processes on morphologies, molar mass distributions, and phase structures of *n*-butyl acrylate–methyl methacrylate copolymers. *J. Appl. Polym. Sci.* 132, 41467 (2015)

[31] Lee, C.F. The properties of core-shell composite polymer latex.: Effect of heating on the morphology and physical properties of PMMA/PS core-shell composite latex and the polymer blends. *Polymer* 41, 1337–1344 (2000)

[32] Carvalho, A.J.F., Job, A.E., Alves, N., Curvelo, A.A.S., Gandini, A. Thermoplastic starch/natural rubber blends. *Carbohydr. Polym.* 53, 95–99 (2003)

[33] Feng, J., Winnik, M.A., Shivers, R.R., Clubb, B. Polymer blend latex films: morphology and transparency. *Macromolecules* 28, 7671–7682 (1995)

[34] Sperling, L.H., Chiu, T.-W., Hartman, C., Thomas, D. Latex Interpenetrating polymer networks. *Int. J. Polym. Mater.* 1, 331–341 (1972)

[35] Sperling, L.H. Interpenetrating polymer networks: an overview. In *Advances in Chemistry*. Klempner, C., Sperling, L.H., Utracki, L.A. (Eds.) (1994), American Chemical Society, Washington DC, 3–38

[36] Lipatov, Y.S. Polymer blends and interpenetrating polymer networks at the interface with solids. *Prog. Polym. Sci.* 27, 1721–1801 (2002)

[37] Sperling, L.H. *Interpenetrating Polymer Networks and Related Materials*. Springer, New York (1981)

[38] Alemán, J. et al. Definitions of terms relating to the structure and processing of sols, gels, networks, and inorganic-organic hybrid materials (IUPAC Recommendations 2007). *Pure Appl. Chem.* 79, 1801–1829 (2007)

[39] Qiao, M., Chen, D., Ma, X., Liu, Y. Injectable biodegradable temperature-responsive PLGA–PEG–PLGA copolymers: Synthesis and effect of copolymer composition on the drug release from the copolymer-based hydrogels. *Int. J. Pharm.* 294, 103–112 (2005)

[40] Mandal, B.B., Kapoor, S., Kundu, S.C. Silk fibroin/polyacrylamide semi-interpenetrating network hydrogels for controlled drug release. *Biomaterials* 30, 2826–2836 (2009)

[41] Bootsma, K. et al. 3D printing of an interpenetrating network hydrogel material with tunable viscoelastic properties. *J. Mech. Behav. Biomed. Mater.* 70, 84–94 (2017)

[42] Lin, F. et al. In situ polymerization approach to cellulose–polyacrylamide interpenetrating network hydrogel with high strength and pH-responsive properties. *Cellulose* 26, 1825–1839 (2019)

[43] Lee, P., Lok, C.-N., Che, C.-M., Kao, W.J. A multifunctional hydrogel delivers gold compound and inhibits human lung cancer xenograft. *Pharm. Res.* 36, 1–10 (2019)

[44] Gersappe, D. et al. The use of graft copolymers to bind immiscible blends. *Science* 265, 1072–1074 (1994)

[45] Zhang, H. et al. Control of selective location of homopolymer-brush grafted nanoparticles in binary polymer blends. *Compos. Sci. Technol.* 200, 108439 (2020)

[46] Robeson, L. Applications of polymer blends: Emphasis on recent advances. *Polym. Eng. Sci.* 24, 587–597 (1984)

[47] Robeson, L. Historical perspective of advances in the science and technology of polymer blends. *Polymers* 6, 1251–1265 (2014)

[48] Paul, D. In *Mechanical Behaviour of Materials VI, Proceedings of the Sixth International Conference, Kyoto, Japan, 29 July–2 August 1991*. Jono, M., Inoue, T. (1992), Pergamon, Oxford, 287–294

[49] Zhang, H. et al. Controlling the enrichment location of brush grafted multi-walled carbon nanotubes at the interface of various polymer blends. *Polymer* 238, 124427 (2022)

[50] Zhang, X. et al. Effects of selective distribution and migration of poly(methyl methacrylate)-grafted nanoclays on the phase behavior of poly(methyl methacrylate)/poly(styrene-co-acrylonitrile) blends. *Polymer*, 124965 (2022)

[51] Bunhu, T., Chaukura, N., Tichagwa, L. Preparation and characterization of polymer-grafted montmorillonite-lignocellulose nanocomposites by in situ intercalative polymerization. *J. Appl. Chem.* 4137398 (2016); DOI: org/10.1155/2016/4137398

[52] Zheng, S., Huang, J., Li, J., Guo, Q. Phase behavior and properties of poly(methyl methacrylate)/poly(vinyl acetate) blends prepared via in situ polymerization. *J. Appl. Polym. Sci.* 69, 675–684 (1998)

[53] Pandey, P. P. Preparation and characterization of polymer nanocomposites. *Soft Nanosci. Lett.* 10, 1 (2020)

[54] Zheng, S., Huang, J., Zhong, Z., He, G., Guo, Q. A polymer of bisphenol A and bisphenol A diglycidyl ether and its blends with poly(styrene-co-acrylonitrile): In situ polymerization preparation, morphology, and mechanical properties. *J. Polym. Sci., Part A: Polym. Chem.* 37, 525–532 (1999)

[55] Zheng, S., Guo, Q., Chan, C. M. Poly(hydroxyether of bisphenol A)/poly(vinyl acetate) blends: In situ polymerization preparation, morphology, and properties. *J. Polym. Sci., Part A: Polym. Chem.* 37, 2329–2337 (1999)

[56] Ogawa, T., Murakami, M. Poly(silmethylene)-based polymer blends. I. In situ polymerization of 1,3-disilacyclobutanes in silicon-based polymers. *J. Polym. Sci., Part A: Polym. Chem.* 35, 399–405 (1997)

[57] Killheffer, D. H. *Banbury the Master Mixer.* Palmerton, New York (1962)

[58] Chen, Z., Wang, Q. Pan-milling mixing – a novel approach to forming polymer blends and controlling their morphology. *Polym. Int.* 50, 966–972 (2001)

[59] Bazli, L. et al. Electrical properties of polymer blend composites based on silicone rubber/EPDM/clay for high voltage insulators. *J. Compos. Compd.* 3, 18–24 (2021)

[60] Utracki, L. History of commercial polymer alloys and blends (from a perspective of the patent literature). *Polym. Eng. Sci.* 35, 2–17 (1995)

[61] Hancock, T. *Personal Narrative of the Origin and Progress of the Caoutchouc or India-Rubber Manufacture in England.* Longman, Brown, Green, Longmans, & Roberts, London (1857)

[62] White, J. L. Development of internal-mixer technology for the rubber industry. *Rubber Chem. Technol.* 65, 527–579 (1992)

[63] Khanra, S., Kumar, A., Ganguly, D., Ghorai, S. K., Chattopadhyay, S. The efficacy of methyl vinyl silicone-g-maleic anhydride in the compatibilization of fluoroelastomer and silicone based super specialty elastomer blend. *J. Polym. Res.* 29, 1–18 (2022)

[64] Wang, J. et al. Dynamic Interfacial Cross-Linking Polymer Blends with Fiber-Like Phase Structure by Recycling. *Macromol. Mater. Eng.* 307, 2100631 (2022)

[65] Kishi, H. et al. Structure formation and conductive properties of epoxy/in situ polymerized methacrylate polymer/silver filler composites. *Polymer* 241, 124520 (2022)

[66] Moran, C. S., Barthelon, A., Pearsall, A., Mittal, V., Dorgan, J. R. Biorenewable blends of polyamide-4,10 and polyamide-6,10. *J. Appl. Polym. Sci.* 133 (2016)

[67] Sengers, W., Van den Berg, O., Wübbenhorst, M., Gotsis, A., Picken, S. Dielectric spectroscopy using dielectric probes: a new approach to study glass transition dynamics in immiscible apolar polymer blends. *Polymer* 46, 6064–6074 (2005)

[68] Tang, Z. et al. Phase transition behavior of poly(vinylidene fluoride) in a blend with poly(butylene adipate) at high temperature. *Polymer* 194, 122409 (2020)

[69] Scobbo Jr, J. Dynamic mechanical analysis of compatibilized polymer blends. *Polym. Test.* 10, 279–290 (1991)

[70] Simon, G. Dielectric relaxation spectroscopy of thermoplastic polymers and blends. *Mater. Forum* 18, 235–264, 1994

[71] Yukioka, S., Inoue, T. Ellipsometric analysis on the in situ reactive compatibilization of immiscible polymer blends. *Polymer* 35, 1182–1186 (1994)

[72] Akoumeh, R., Elzein, T., Martínez-Campos, E., Reviriego, F., Rodríguez-Hernández, J. Fabrication of porous films from immiscible polymer blends: Role of the surface structure on the cell adhesion. *Polym. Test.* 91, 106797 (2020)

[73] Thareja, P., Velankar, S. Particle-induced bridging in immiscible polymer blends. *Rheol. Acta* 46, 405–412 (2007)

[74] Liu, J., Jean, Y., Yang, H. Free-volume hole properties of polymer blends probed by positron annihilation spectroscopy: miscibility. *Macromolecules* 28, 5774–5779 (1995)

[75] Ranganathaiah, C. Positron annihilation spectroscopy: polymer blends and miscibility. *Charact. Polym. Blends* 877–920 (2014)

[76] Smith, C. G. et al. Analysis of synthetic polymers and rubbers. Anal. Chem. 65, 217–243 (1993)

[77] *Characterization of Polymer Blends: Miscibility, Morphology and Interfaces.* Thomas, S., Grohens, Y., Jyotishkumar, P. (Eds.) (2014), Wiley-VCH, Weinheim

[78] Smentkowski, V. et al. Using time-of-flight secondary ion mass spectrometry and multivariate statistical analysis to detect and image octabenzyl-polyhedral oligomeric silsesquioxane in polycarbonate. *Appl. Surf. Sci.* 253, 1015–1022 (2006)

[79] Ovchinnikova, O. S. et al. Helium ion microscopy with secondary ion mass spectrometry for nanoscale chemical imaging and analysis of polyolefins. *ACS Appl. Polym. Mater.* 3, 3478–3484 (2021)

[80] Prasad, A., Salim, N. V., Mozetič, M., Kailas, L., Thomas, S. Time-of-flight secondary ion mass spectrometric analysis of polymer surfaces: A review. *J. Appl. Polym. Sci.*, 52286 (2022)

[81] Inoue, T., Ougizawa, T., Yasuda, O., Miyasaka, K. Development of modulated structure during solution casting of polymer blends. *Macromolecules* 18, 57–63 (1985)

[82] Jo, K.-I. et al. Enhanced dynamics of confined polymers near the immiscible polymer–polymer interface: Neutron reflectivity studies. *ACS Macro Lett.* 9, 210–215 (2020)

[83] Dong, J., Ozaki, Y. FTIR and FT-Raman studies of partially miscible poly(methyl methacrylate)/poly(4-vinylphenol) blends in solid states. *Macromolecules* 30, 286–292 (1997)

[84] Lipatov, Y. S., Shumsky, V., Getmanchuk, I., Gorbatenko, A. Rheology of polymer blends. *Rheol. Acta* 21, 270–279 (1982)

[85] Gómez, C. M., Figueruelo, J. E., Campos, A. Thermodynamics of a polymer blend solution system studied by gel permeation chromatography and viscosity. *Macromol. Chem. Phys.* 200, 246–255 (1999)

6 Experimental Approaches for the Evaluation of Phase Separation Process in Polymer Blends

6.1 Introduction

As mentioned in **Chapter 2**, phase separation of polymer blends incorporates thermodynamics, morphology, and solid physical properties [1]. The morphology of miscible or immiscible polymer blends can be observed by microscopy techniques such as optical microscope (OM), scanning electron microscope (SEM), and transmission electron microscope (TEM). The thermal and mechanical properties in binary components are determined by their physical interaction. Binary polymer blends on account of limited reactive functional groups result in phase separation, and, hence, poor interfacial adhesion and poor mechanical properties. These limitations can be avoided by adding compatibilizers. One property that reflects compatibility effectiveness is the glass transition temperature (T_g) and it is measured by different techniques such as dynamic mechanical analysis (DMA), differential scanning calorimetry (DSC), and rheological method. Based on the DSC and DMA results, a single T_g is referred to as miscible, and two distinct T_gs are immiscible or phase separated. However, a single T_g on DSC does not necessarily mean the binary components are miscible. In this chapter, the effectiveness of various techniques, their limitations and advantages, in the determination of the level of miscibility in polymer blends is discussed.

6.2 Evaluation of Phase Separation Behavior in Polymer Blends by DSC

Both DSC and modulated DSC (MDSC) techniques can be employed to characterize the phase behavior of polymer blends. The miscibility of polymer blends is described by the behavior of the T_g. The T_g reflects the molecular rearrangement, transition, and relaxations in polymer blends. Measuring T_g to determine whether polymer blends are miscible or immiscible is relatively simple; if a polymer blend of two individual polymers shows a single T_g then they are miscible while two T_gs indicates an immiscible polymer blend. Based on the relationships between composition and weight fraction of individual polymer blends, a number of theoretical equations have been established.

Fox equation (Equation 6.1) [2]:

$$\frac{1}{T_g} = \frac{w_1}{T_{g1}} + \frac{w_2}{T_{g2}} \tag{6.1}$$

Where T_{g1} and T_{g2} are the glass transitions of polymer 1 and polymer 2, respectively, and w_1 and w_2 are the respective weight fractions. This equation holds for polymer pairs with no specific interactions.

Gordon–Taylor equation (Equation 6.2) [3]:

$$T_g = \frac{w_1 T_{g1} + K w_2 T_{g2}}{w_1 + K w_2} \tag{6.2}$$

Kwei equation (Equation 6.3) [4]:

$$T_g = \frac{w_1 T_{g1} + K w_2 T_{g2}}{w_1 + K w_2} + q w_1 w_2 \tag{6.3}$$

Where q is a parameter that incorporates the hydrogen bonding strength.

Zuideveld et al. [5] employed DSC and DMA to assess the miscibility of PLLA and PLLA-based co-polyesters (hb-PLLA) blends. Different compositions of PLLA/hb-PLLA (90/10, 70/30, and 50/50 wt.%) blends were prepared via melt mixing at 190 °C. At this composition range, all the blends showed single T_g, indicating miscibility. They also used the Fox equation (Equation 6.1) to compare the T_g and found a small positive deviation of less than 2. Na et al. [6] used DSC to evaluate the effects of the addition of different copolymers PCL-b-monomethoxy-PEG (PCL-b-MPEG) and (poly(ε-caprolactone)-b-poly(ethylene glycol) (PCL-b-PEG) on the miscibility and thermal behavior of PLA/PCL blends. For the thermal study, they used two different forms of PLA (poly(L-lactide) (PLLA) and poly(DL-lactide) (PDLLA)). Thin films with various compositions (PLLA/PCL, PLLA/PCL-b-PEG, PDLLA/PCL, PDLLA/PCL-b-PEG) were prepared via solution casting under chloroform, followed by compression molding. From the DSC thermograms, they found the T_g of PLLA decreased from 61 °C to 38 °C when 50 wt.% PCL-b-PEG was added; however, the T_g of PCL-b-PEG remained the same as the T_g of PCL. Based on the T_g it is immiscible [7–9]; however, Sheth et al. reported that PLLA is miscible with PEG [10]. Here, the decrease in T_g after the addition of PCL-b-PEG is explained due to the mixing of PEG chains to the amorphous PLLA confirming the compatibilizing effect of PCL-b-PEG. The miscibility of binary blends is influenced by chemical interactions, such as dipole-dipole and hydrogen bonding [11]. Based on the evaluation of solubility parameters, amorphous PDLA is miscible with PMA since the difference δ is less than 0.5 and it is less likely that it is miscible with PMMA as the δ values difference between PDLA and PMMA is greater than 0.5. Eguiburu et al. [12] studied the impact of blending amorphous PDLA and semicrystalline PLLA with PMA and PMMA on their miscibility and thermal properties. From DSC, they found that a single T_g was observed for both the blends (PDLA/PMA and PDLA/PMMA), conforming to miscibility. This experimental observation was validated by using the Gordon–Taylor equation (Equation 6.2). Furthermore, they investigated the effect of blending semicrystalline/amorphous polymers (PLLA/PMMA and PLLA/PMA). Both the blends were miscible in the melt state and phase separated when cooled from melt due to the crystallization of PLLA. Zhang et al. [13] studied the effect of different processing methods on the thermal and miscibility properties of PDLA/PMMA blends. The blends were prepared via solution-casting and solution/precipitation. Samples prepared by solution-precipitation showed a single T_g and those prepared by solution casting showed two T_gs conforming to immiscibility (consistent with solubility parameters rule).

The DSC technique has limits when the difference between the glass transition temperatures of polymers involved in the blend is less than 20 °C.

6.3 Evaluation of Phase Separation Behavior in Polymer Blends by DMA

DMA is the most popular technique used to measure viscoelastic properties and T_g of polymeric materials. It is a frequency response analysis under tensile or compression mode. It provides information about the storage modulus (E') and loss modulus (E'') of the materials. The ratio of E'' to E' is called the tan(delta) or loss tangent. The T_g is considered the temperature at which the loss tangent is maximum [14]. Hence, besides DSC, DMA can be used to detect the T_g of polymeric materials and is more sensitive in evaluating the miscibility [15]. As a criterion, the existence of a single or two T_gs has been widely used to predict the miscibility of binary components [16]. Miscible polymeric blends show a single T_g, while immiscible polymeric blends show two distinct T_gs.

PLLA is a biodegradable polymer used for biomedical applications. However, its brittleness and low HDT (heat distortion temperature) limit its range of applications [17, 18]. Hence, PLLA is blended polymers, such as PCL [19], PBS [20, 21], and PU [22], to improve its properties. For example, Qiu et al. [23] blended PLLA with poly(oxymethylene) (POM) and studied their miscibility and mechanical properties by using DMA. PLLA/POM blends were prepared via melt mixing then hot-pressed at 190 °C and quenched to room temperature. From the results, PLLA/POM blends are miscible in the melt at 190 °C due to a specific interaction, which was confirmed by FTIR to be between –C=O and C–H groups of PLLA and POM, respectively, and the phases separated (crystallization-induced phase separation due crystallized component POM) when quenched to room temperature as confirmed by the existence of two T_gs shown in Figure 6.1. Further, a depression in T_gs of both POM and PLLA was observed. This was explained by the fact that when POM crystallizes it creates a negative pressure on PLLA which results in shrinkage. The negative pressure at the interface between the two components exhibits a reduction in free volume resulting in depression in the T_gs of the individual components. The decreased T_gs of the PLLA/POM blends resulted in increased in ductility as compared to neat PLLA and POM. Improved ductility of the PLLA/POM blends, as compared with neat PLLA and POM, was also observed, which was attributed to higher molecular mobility due to the T_g depression for both PLLA and POM in the blends.

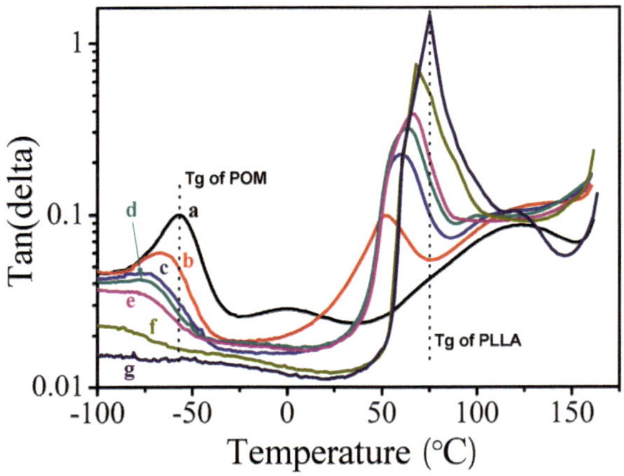

Figure 6.1 Dynamic loss for the PLLA/POM blends as a function of temperature: (a) neat POM, (b) PLLA/POM = 20/80, (c) PLLA/POM = 40/60, (d) PLLA/POM = 50/50, (e) PLLA/POM = 60/40, (f) PLLA/POM = 80/20, and (g) neat PLLA. Reproduced with permission from [23]. Copyright 2013, American Chemical Society

Poly(tetramethylene adipate-co-terephthalate) (PTAT) displays higher hydrophilicity and easier processability as compared to PLLA. Liu et al. [24] blended PLLA with PTAT to improve the mechanical properties of PLLA. The blends with varying ratios of PLLA and PTAT were prepared via solution casting under chloroform and the miscibility and thermal and mechanical properties were examined by DSC, AFM, and DMA. The variation in miscibility with the variation of the content of PTAT was obtained. Among them, PLLA/PTAT (75/25 wt.%) blend indicated partial miscibility, as depicted in Figure 6.2, and exhibited higher miscibility and higher tensile strength than other blend ratios. This miscibility PLLA/PTAT (75/25 wt.%) blend demonstrated improved tensile strength as compared to other ratios.

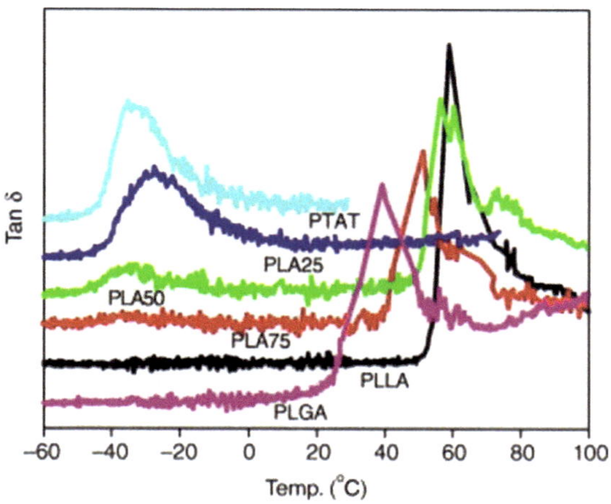

Figure 6.2 Tan δ curves of pure and blend membranes measured by DMA. Reproduced with permission from [24]. Copyright 2005, Elsevier Science Ltd

The parameters controlling the phase separation during crystallization-induced phase separation in PLLA/PBA blends were examined by Woo et al. [25] based on time maintained at the highest melting point and entanglement density. The PLLA/PBA blends were prepared via solution casting under chloroform. For 50/50 wt.% blend, as schematically represented in Figure 6.3, they were able to attain phase-separated morphology by maintaining the sample at the melting point for a shorter time, in this case for 2 min. In contrast, maintaining the sample for 10 min at the melting point resulted in a homogenous (no phase separation) domain. In the case of the blend melted for 2 min (short time), this results in lower entanglement density; accordingly, segregation of the PBA component chains from the crystallizable PLLA during crystallization would be simpler, hence this separated PLLA and PBA domains. However, for 10 min holding time, there is greater chain entanglement density, and it is difficult to segregate PBA chains from PLLA. They concluded that for 2 min holding time, lower chain entanglement density resulted in higher nucleation density, small crystalline domains, and fast growth rate. However, for 10 min holding time, higher chain entanglement density resulted in lower nucleation density and slow growth rate of PLLA spherulite.

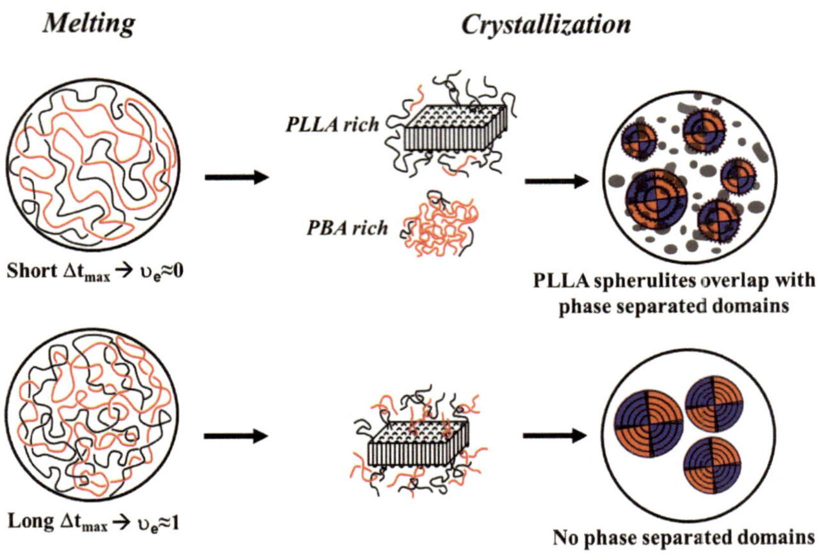

Figure 6.3 Scheme showing the effect of Δt_{max} on the formation of phase separation in PLLA/PBA (50/50) blend system. Reproduced with permission from [25]. Copyright 2012, American Chemical Society

Komalan et al. [26] used DMA techniques to study the compatibility of nylon copolymer (PA6,66) and ethylene propylene diene (EPDM) blends. Sample composition of 70/30 wt.% PA6,66/EPDM blends with various concentrations of compatibilizer (EPM-g-MA) were prepared. The results revealed that all the blends without compatibilizer exhibited two distinct T_gs confirming immiscibility of the binary components. Similarly, two distinct T_gs were observed after the addition of compatibilizer. This indicated that the addition of compatibilizer showed no significant change in the miscibility of the blends. The compatibilizer at 2.5 wt.% indicated maximum peak (tan δ) due to interfacial interaction.

■ 6.4 Evaluation of Phase Separation Behavior in Polymer Blends by Rheology

Another common technique employed to evaluate the phase behavior of binary polymeric material is rheological measurements. Rheology measures the viscoelastic properties of materials in relation to their microstructure. It provides information about storage modulus (G'), loss modulus (G''), and tan δ, which is the ratio of G''/G'. One of the most popular rheological fingerprints used to determine the phase behavior of polymer blends is frequency sweep. In this protocol, the G' is varied as a function of frequency at a fixed strain (in the region of LVR) and fixed temperature. When the phase separation occurs, compared to the G'', the change is more apparent in the G'. During phase separation, in the frequency sweep protocol, a shoulder in the G' or deviation from the scaling law ($G' \sim \omega^2$) is observed at lower frequencies. However, there is no deviation in the scaling law for miscible binary polymer blends. Overall, in the regime of phase separation, more elastic, longer relaxation time, and a failure of the time-temperature superposition (TTS) are observed. Apart from phase behavior, the frequency sweep protocol provides information about the phase morphology of phase-separated polymer blends, whether it is droplet matrix or co-continuous morphologies. Another approach to determine the phase separation behavior is by plotting η' (imaginary viscosity) as a function η'' (real viscosity), referred to as Cole–Cole plots. For miscible polymer blends, the graph displays a semi-circular arc, and when the binary polymer blend is phase separated, the graph reveals a semi-circular arc with a drifting tail. This additional tail is due to additional relaxation attributable to phase-separated domains.

It should be noticed that the phase separation temperature determined by using frequency sweep, temperature ramp, and time sweep are all affected by the time of experiments. The phase separation temperature determined by the above rheological methods is accurate only when the phase separation is faster than the measurement speed. Otherwise, there must be some retardant effects in the rheological determined phase separation temperature, which can be regarded as apparent values.

Polios et al. [27] employed rheology to examine the phase behavior of PS/PVME (20/80 wt.%) blends prepared by solution casting under toluene. The sample was held at 124 °C for a long interval of time (42 h). At the start of phase separation, an increase in G' at low frequencies was observed owing to the interconnected structure of the phase-separated component. Further increase in annealing time exhibited a decrease in G' due to loss of connectivity between the binary phases attributed to the coarsening and breakup of the heterogeneous domains [27]. The miscibility of SMA/PMMA blends depends on the amount of maleic anhydride (MA), and solvent used [28]. Chopra et al. [29] investigated the effects of maleic anhydride content on the rheology and phase behavior of PMMA/SMA blends. In order to detect the phase separation temperature, a dynamic temperature sweep was conducted on PMMA/SMA blends with varying amounts of MA content (8, 14, and 32 wt.%) in the SMA component at a heating rate of 1.5 °C/min. The onset of phase separation temperature was taken at the change in slope of storage modulus (G'), and the maximum peak in tan δ was taken as the onset of phase separation temperature. The authors found the phase separation temperature for PMMA/SMA (50/50 wt.%) with 14 wt% MA content was around 245 °C. This was consistent with the Cole–Cole plot result. With a similar approach, Kapnistos et al. [30] examined the phase behavior of PS/PVME (40/60) blends, as shown in Figure 6.4.

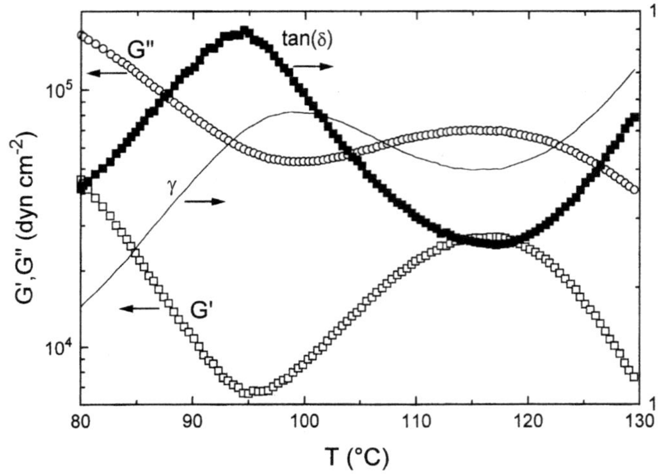

Figure 6.4 Isochronal (ω = 0.1 rad/s) dynamic temperature ramp of 40/60 PS/PVME for an applied stress amplitude of 1000 dyn/cm^2 and a heating rate of 0.1 °C/min: G', □; G'', ○; tan δ, ■; γ, –. Reproduced with permission from [30]. Copyright 1996. American Chemical Society

■ 6.5 Evaluation of Phase Separation Behavior in Polymer Blends by Broadband Dielectric Relaxation Spectroscopy

Broadband dielectric relaxation spectroscopy (DRS) is used to determine the dielectric response of materials upon application of an electric field. The primary criteria to measure the dielectric properties of a material is whether the material should be dielectric active or have dipole moments. The DRS provides information about the dielectric constant (ε'), dielectric loss or loss factor (ε''), and tan δ, which is the ratio of dielectric loss to the dielectric constant ($\varepsilon''/\varepsilon'$). These dielectric properties of polymeric materials are analogous to mechanical properties. Hence, DRS can be used to detect the glass transition temperature. In the context of the glass transition temperature, rheological measurements are insensitive for the detection of viscoelastic properties of blends when their difference in T_g is very small (weak dynamically asymmetric blends). Here, the dielectric measurement is more sensitive to evaluate the various relaxation processes in weakly dynamic asymmetric blends. The mechanical and electrical response of polymeric materials is influenced by chain dynamics or various relaxations. As a consequence, understanding polymer chain dynamics via DRS plays an important role in many applications. Sasabe et al. [31] employed DRS to investigate the structural dynamics of PVDF. On the frequency versus dielectric loss curve at various temperatures, the authors found three relaxations, namely α_a, α_c, and β relaxations which are assigned to molecular chain motion in the amorphous region, molecular chain motion in the crystalline region, and local motion, respectively. These relaxations were explained schematically for pure PMMA by Sengwa et al., as presented in Figure 6.5 [32]. Accordingly, α relaxation is due to segmental motion and is found at high temperatures and low frequencies. The β is a localized motion of $COOCH_3$ chains and is located at a lower temperature compared to α relaxation. Near the T_g the α and β relaxations merge and become αβ relaxation [33].

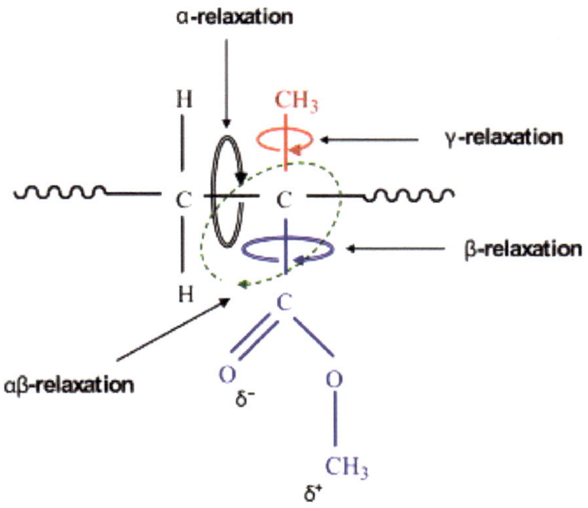

Figure 6.5
Schematic representation of various relaxation processes associated with PMMA dynamics. Reproduced with permission from [32]. Copyright 2014, Elsevier Science Ltd

It is well known that PVDF is miscible with PMMA in the melt owing to the dipole-dipole interaction. The key factors, such as blend composition, cooperative length scale, and confinement, influencing the structural relaxation dynamics in PVDF/PMMA blends with different compositions were investigated extensively by Sy and Mijovic using DRS [34]. Yang et al. [35] evaluated the impact of GO on the structural dynamic relaxations of PVDF/PMMA blends. The composite films, which were coded FM-x (FM= 50/50 PVDF/PMMA, x = 0.2, 0.5, and 1 wt.%), were prepared solution casting under DMF followed by compression molding. As depicted in Figure 6.6, at high temperatures (140 to 170 °C) for the blend without GO, prominent α_a relaxation was observed as it is the segmental motion of the amorphous phase. On the other hand, when cooled from the melt owing to the crystallization of PVDF, two relaxations, α_c and α_m (due to miscibility), were obtained. The α_m relaxation is more pronounced with increased GO content, conforming to improved miscibility in the amorphous phase. It is well noted as well that at high loading of GO, the α_c relaxation shifted to higher frequencies indicating imperfect crystals. They concluded that incorporation of GO augments miscibility and impedes crystallization of the blend [35].

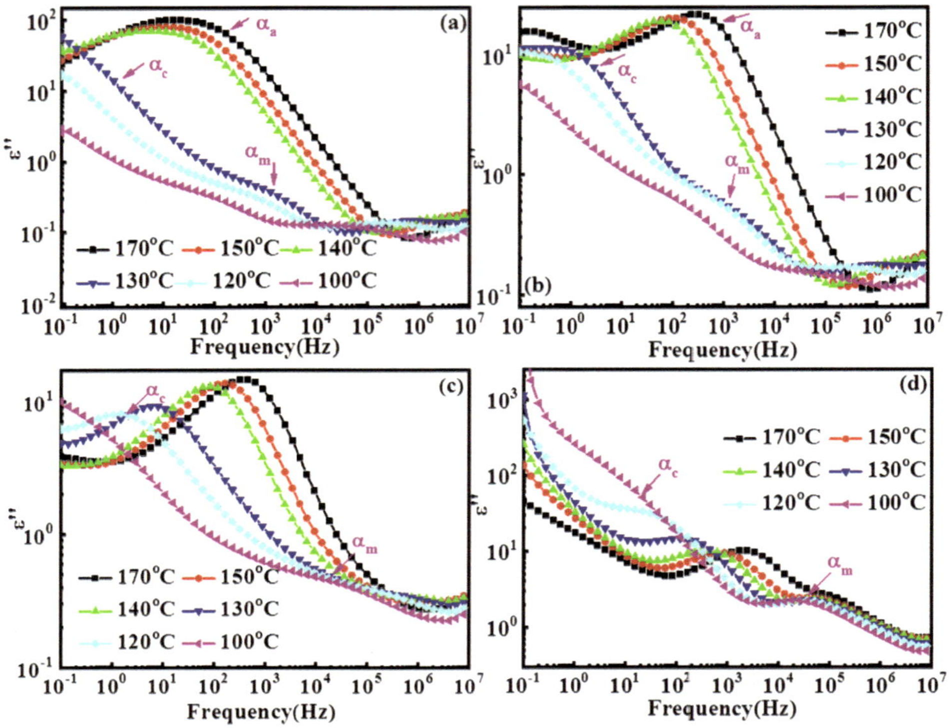

Figure 6.6 Relaxation spectrum under different temperature of (a) FM, (b) FM-0.2, (c) FM-0.5, and (d) FM-1. Reproduced with permission from [35]. Copyright 2021, Elsevier Science Ltd

Another interesting study was reported by Madbouly on the phase behavior of PMMA/PαMSAN (poly(α-methylstyrene-co-acrylonitrile)) blends [36]. As mentioned above, α relaxation is related to the glass transition temperature. A single α relaxation in all PMMA/PαMSAN blend compositions was observed, indicating miscibility of the blends at the segmental level. However, the β relaxation of the blends was independent of the blend composition hence no miscibility.

Zhang et al. [37] fabricated a thin film of PVPh/PVEE (poly(4-vinylphenol)/poly(vinyl ethyl ether)) with PVPh composition of 10 to 50 wt.% via solution casting and analyzed the structural relaxation and miscibility using DRS and DSC. From DSC, a single T_g was observed for all the blend compositions. From DRS, on the one hand, two α relaxations were achieved at a lower amount of PVPh less than 30 wt.%. The two relaxations were due to the hydrogen-bonded segment of the binary blends and unbonded PVEE segments. On the other hand, a single α relaxation at PVPh greater than 30 wt.% was observed due to increased hydrogen-bonded segments as the content of PVPh increases. The higher the hydrogen bonding between the binary blends increased the intermolecular coupling, resulting in single relaxation [37].

Jin et al. [38] studied the influence of ageing on the dielectric relaxation of PMMA/PEO blends using DRS. The results revealed that after ageing for one day at 25 °C, a new β-relaxation peak was observed and was explained due to partially ordered PEO chains. Sengers et al. [39] also used DRS to evaluate the effect of polarity on the glass transition behavior of PS/PE blends. PS/PE blends are immiscible, as evidenced by two distinct α relaxations associated with the T_g of individual polymers. As shown in Figure 6.7, the T_g dynamics are not influenced by blend concentration or morphologies [39].

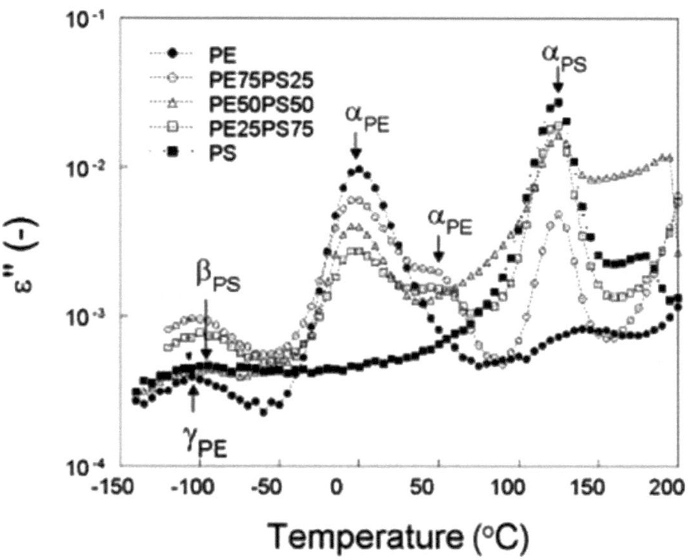

Figure 6.7 Temperature dependence of the dielectric loss at 1 kHz for PS/PE blends. The arrows indicate the position of the intracrystalline relaxation process in PE (α_{PE}), the glass transition processes of PS (α_{PS}) and PE (β_{PE}), and the sub-glass transition processes of PS (β_{PS}) and PE (γ_{PE}). Reproduced with permission from [39]. Copyright 2005, Elsevier Science Ltd

■ 6.6 Evaluation of Phase Separation Behavior in Polymer Blends by Transmission Electron Microscope

Microscopy techniques, such as optimal microscope (OM), transmission electron microscope (TEM), and scanning electron microscope (SEM), are used to determine the phase behavior and morphology of binary polymer blends. Among them, TEM is a powerful technique that provides information about nanoparticle morphology, shape, size, and location in different phases. Before the instrumental procedure, sample preparation involves staining, etching, sectioning/microtoming, fracturing, etc. One key challenge of using TEM is that it offers information on a very small area of the sample; hence it is necessary to examine various areas in order to get comprehensive information about the specimen. The miscibility of poly(methyl acrylate) (PMA) and hydroxy-modified polystyrene (PS(OH)) was studied by using TEM [40]. For the TEM, samples were prepared by solution casting under toluene. The thin film was imbedded in epoxy and cured at 60 °C for sectioning. It was found that the phase behavior of PMA/PS(OH) depends on the hydroxy content present in PS(OH). Lower hydroxy content in PS(OH) showed phase separation and altered to a single phase when the hydroxy content increased [40]. Tiwari et al. [41] explored the impact of the presence of nanoclay (organically modified) (MMT) and compatibilizer PP-g-MA on the phase morphology of PP/PS blends fabricated by melt mixing. PP with different molecular weights were used for the experiment; that is, L-PP, M-PP, and H-PP were assigned to low M_w, medium M_w, and high M_w of PP, respectively. As seen from TEM micrographs reported in Figure 6.8, the MMT particles are localized at the PP phase and interface when the content of PS in the blend was 30 wt%. MMT was localized at the PS phase when the PS content in the blend was 90 wt%. It was also found that as the molecular weight of PP increases, the droplet of PS morphology becomes elongated due to shear force. A reduction in dispersed PS particle size was pronounced when PS was blended with H-PP [41].

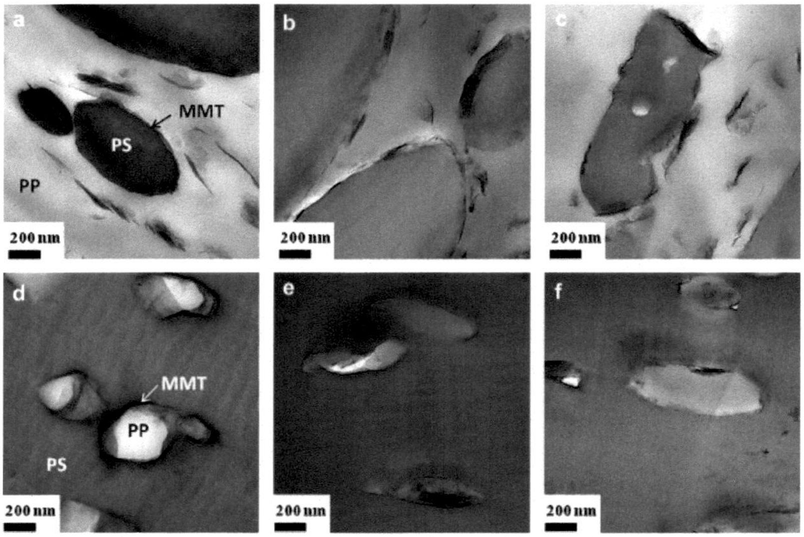

Figure 6.8 Transmission electron microscope micrographs showing location of the MMT particles in PP/PP-g-MA/MMT/PS blend having 30 wt.% PS (top row) and 90 wt.% PS (bottom row). The blends were prepared with different PP grades: L-PP (a and d), M-PP (b and e), and H-PP (c and f). Images were taken from the core and viewed perpendicular to the flow direction (FD). Reproduced with permission from [41]. Copyright 2011, Elsevier Science Ltd

6.7 Evaluation of Phase Separation Behavior in Polymer Blends by Scanning Electron Microscope

To recap, the blending of two polymers provides several different morphologies. The type of morphologies created depends on the processing methods [42, 43]. SEM techniques are normally employed to determine images of the morphology. Sample preparation involves cryofracture by liquid nitrogen, etching one of the phases, and coating with gold or carbon. Galloway et al. [44] investigated the impact of the addition of different molecular weight copolymers PS-PE copolymers on the morphology of PS/HDPE blends prepared by melt processing. The molecular weight of the PS-PE copolymers was 6, 40, 100, and 200 kg/mol. The SEM thermographs shown in Figure 6.9 show that the 50/50 neat PS/HDPE blends showed co-continuous morphology [44]. Among all the molecular weights of copolymers, 40 kg/mol PS-PE copolymer gave a significant reduction in domain size and was the most effective in reducing coarsening.

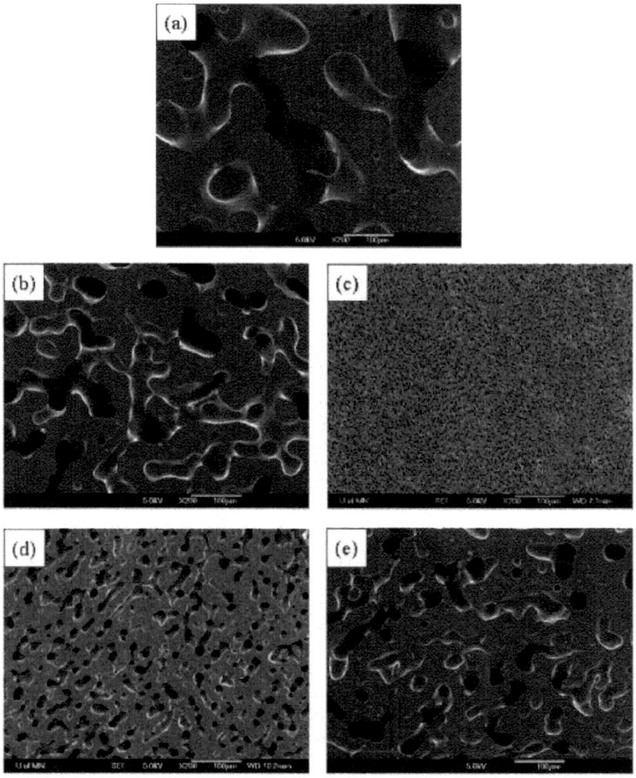

Figure 6.9
SEM micrographs of 50/50 PS/PE blends with 1% PS-PE after mixing at 50 rpm for 10 min at 170 °C followed by annealing for 15 min:
(a) no block copolymer,
(b) 6 kg/mol PS-PE,
(c) 40 kg/mol PS-PE,
(d) 100 kg/mol PS-PE, and
(e) 200 kg/mol PS-PE.
(Scale bar = 100 µm).
Reproduced with permission from [44]. Copyright 2005, Elsevier Science Ltd

■ 6.8 Evaluation of Phase Separation Behavior in Polymer Blends by Optical Microscope

Another popular method to study the crystalline morphology of polymer blends is a polarized optical microscope (POM) equipped with a hot stage. POM is advantageous because relatively easy to use, and it directly provides the morphology and flow properties of solid or liquid polymeric materials with the application of shear or without shear. The system consists of a precise temperature controller and, hence, it is accurate for studying LCST and UCST polymer blends by varying the heating and cooling rate. For example, Xavier et al. [45] employed POM to study the phase behavior of PS/PVME blends. The samples were heated from 80 °C to 180 °C at a heating rate of 10 K/min. The phase separation temperature was obtained at 122 °C for 10/90 blends [45]. Gebrekrstos et al. [46] explored the influence of shear on the crystalline phase, and morphology in PVDF by POM. As depicted in Figure 6.10, samples were heated to 220 °C, kept for 3 min to remove thermal history and cooled at 10 K/min. Shear was applied at different temperatures. The results revealed that a higher rate of crystallization occurs with increasing shear rate.

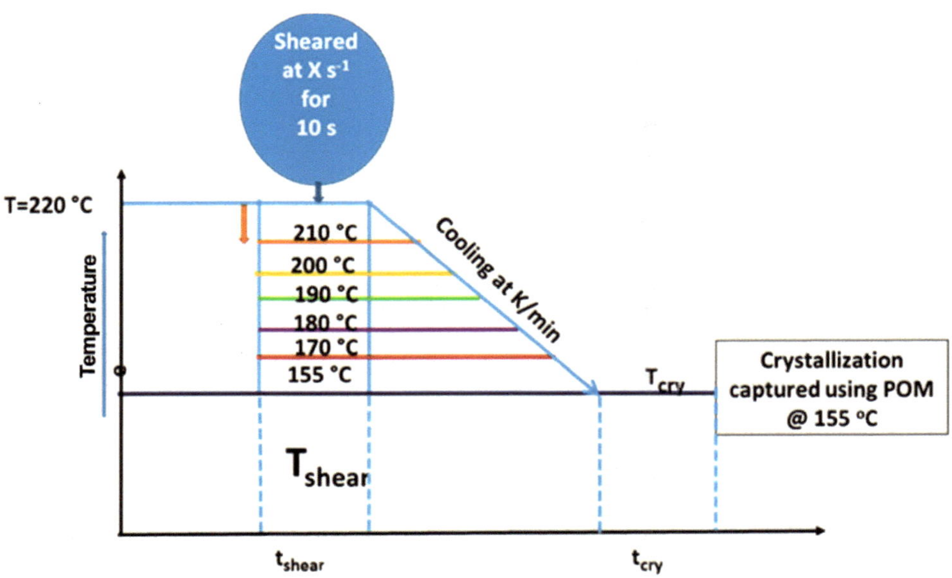

Figure 6.10 Experimental procedure for samples crystalline morphology captured at 155 °C with x = 1, 5, 10, 20, 30, and 40. Reproduced with permission from [46]. Copyright 2016, American Chemical Society

It is reported that PVDF/PBS blend is miscible in the melt, but phase separated when cooled from the melt due to crystallization [47]. Mao et al. [48] used the concept of the crystallization-induced phase separation approach to study the selective localization of CNTs in PBS/PVDF blends (the samples were designed by P for PBS and F for PVDF (PF)). POM was used to observe the crystalline morphology during cooling from the melt. As depicted in Figure 6.11, PVDF spherulitic morphology was observed in the sample crystallized isothermally for 20 min. When further cooled to 90 °C, PBS crystallites are observed in the interspherulitic region of PVDF. After adding CNTs, ring banded PVDF spherulites are achieved; however, the rate of crystallization remained unchanged.

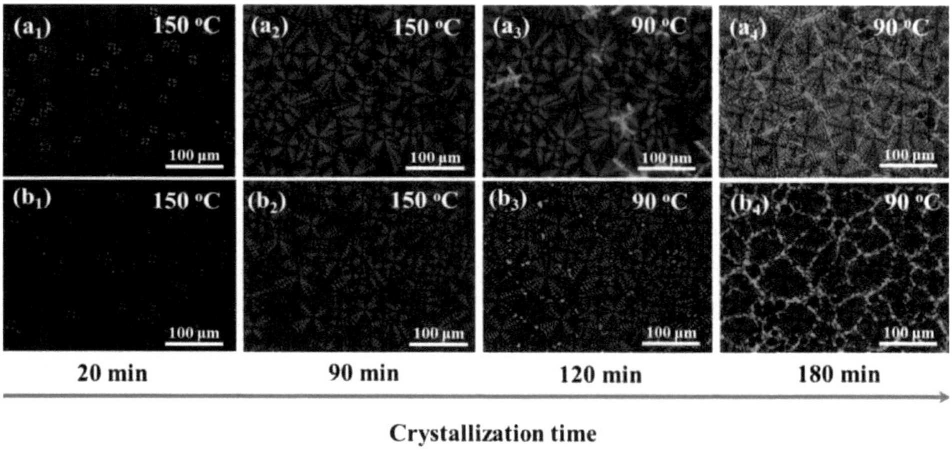

Figure 6.11 POM images of nanocomposites obtained by isothermal crystallization: (a_1, a_2) for P5F5 sample crystallization at 150 °C; (a_3, a_4) for P5F5 crystallization at 90 °C; (b_1, b_2) for P5F50.5 crystallization at 150 °C; and (b_3, b_4) for P5F50.5 crystallization at 90 °C. Reproduced with permission from [48]. Copyright 2020, American Chemical Society

The crystallization-induced phase separation method can be utilized as a useful way to control the structures of crystallizable polymer blends. Here when PVDF is crystallized first, the PBS lamellae are sited in the interspherulitic region of PVDF. Using step-by-step crystallization, better dispersion of CNT and PBS in PVDF was observed; as a result, a high dielectric permittivity value was achieved. The general mechanism is explained schematically in Figure 6.12 [48].

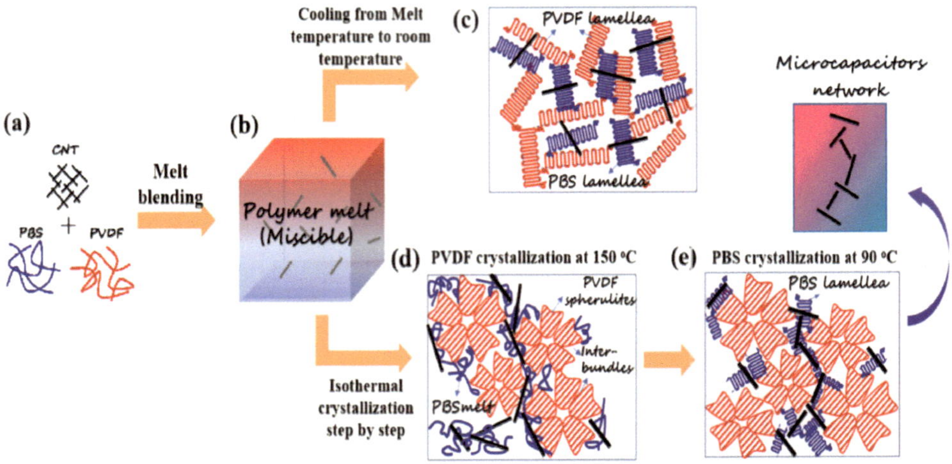

Figure 6.12 Cartoon representations showing the preparation and processing to tailor the distribution of CNTs in a PVDF/PBS miscible blend via crystallization. (a) Fabrication of nanocomposites through the melt blending method, (b) CNTs exhibit good dispersion in the miscible polymer melt, (c) PBS crystallites and CNTs exhibit good dispersion in the PVDF component of the nanocomposites treated through quick cooling, (d) dispersion of the PBS melt and CNTs in the interspherulites of the PVDF component during isothermal crystallization at 150 °C, and (e) dispersion of PBS crystallites and CNTs after isothermal crystallization at 90 °C. Reproduced with permission from [48]. Copyright 2020, American Chemical Society

6.9 Evaluation of Phase Separation Behavior in Polymer Blends by Atomic Force Microscope

The surface morphology of polymeric material can be determined by atomic force microscopy (AFM) due to the modest requirements for sample preparation. AFM is advantageous over SEM and TEM because AFM provides information on the nanometric scale of thick samples [49–51]. As the focus is phase separation detection in polymer blends, a single phase detected by DMA and DSC measurements does not always mean that the blends are necessary miscible. For instance, in the temperature or composition series between the spinodal and binodal points, miscible blends are metastable in relation to concentration fluctuations which are difficult to visualize in DSC or DMA [52, 53]. Wen et al. [54] used AFM to evaluate the surface phase behavior of PMMA/SAN blends annealed at different temperatures. They found that the phase separation mechanism occurred by spinodal decomposition. The impact of composition on the miscibility of PLA/PBAT blends was studied by using AFM [55]. Dispersed phase morphology of L-PBAT at 10 to 30 vol.% composition was observed and became co-continuous at 40 to 50 vol.% as shown in Figure 6.13. Phase inversion was obtained at 70 vol.% of L-PBAT. As a result of phase separation, other studies by AFM on PS/poly(*n*-butyl methacrylate) (PBMA) showed that by varying the composition and solvent island/ribbon type of surface morphology was obtained [56]. Liu et al. [57] fabricated thin films via spin coating under dichloromethane and analyzed the phase behavior by AFM in PEI/PCL blends. They found that the surface morphology of the binary blends depends on the rate of solvent evaporation and composition of the blends.

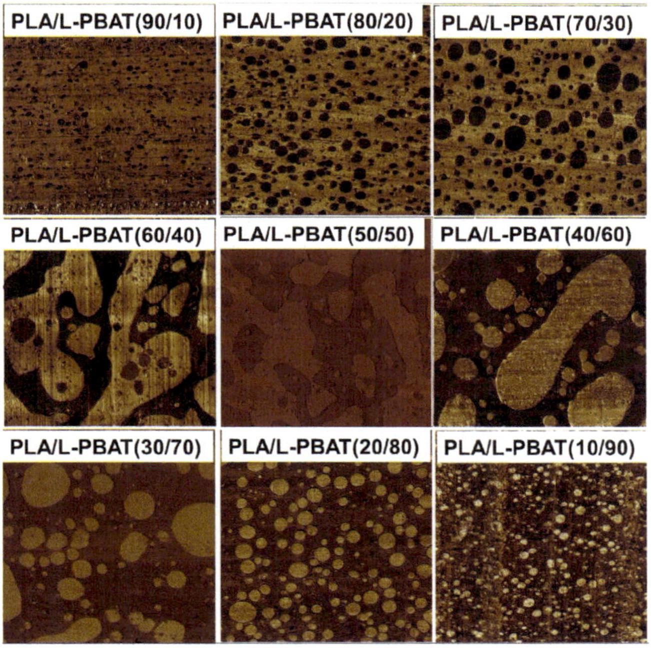

Figure 6.13 AFM images of PLA/L-PBAT blends at different blend compositions. The darker phase is the L-PBAT phase, and the brighter phase is the PLA phase in all images. All the micrographs are 30 μm × 30 μm. Reproduced with permission from [55]. Copyright 2015, Elsevier Science Ltd

Bar et al. [58] deposited a-PP/SEBS and i-PP/SEBS blends at different temperatures, such as 130, 170, and 190 °C, under xylene. The impact of temperature on the phase microstructure of the blends was examined. The results revealed that at lower content of SEBS 20 wt.%, small and uniform well-dispersed morphology was obtained. Upon increasing the concentration of SEBS, large domain sizes and non-uniform phase-separated domains were observed, as shown in Figure 6.14 [58]. Cabral et al. [59] detected the spinodal decomposition phase separation pattern of tetramethyl bisphenol A polycarbonate (TMPC) and polystyrene (PS) TMPC/PS 50/50 blend thick films by using AFM. Prosycevas et al. [60] prepared PS/PMMA blends with 25/75 up to 75/25 wt.%/vol.% (w/v) composition via the solvent casting method under toluene and analyzed their morphologies by AFM. When the PMMA bulk concentration is greater than 50%, phase-separated morphologies were obtained. This was attributed to the differing solubility of the polymers in the solvent. Therefore, it can be concluded that, as compared to the other microscopic techniques, AFM provides information from surface to bulk with better resolution and morphology.

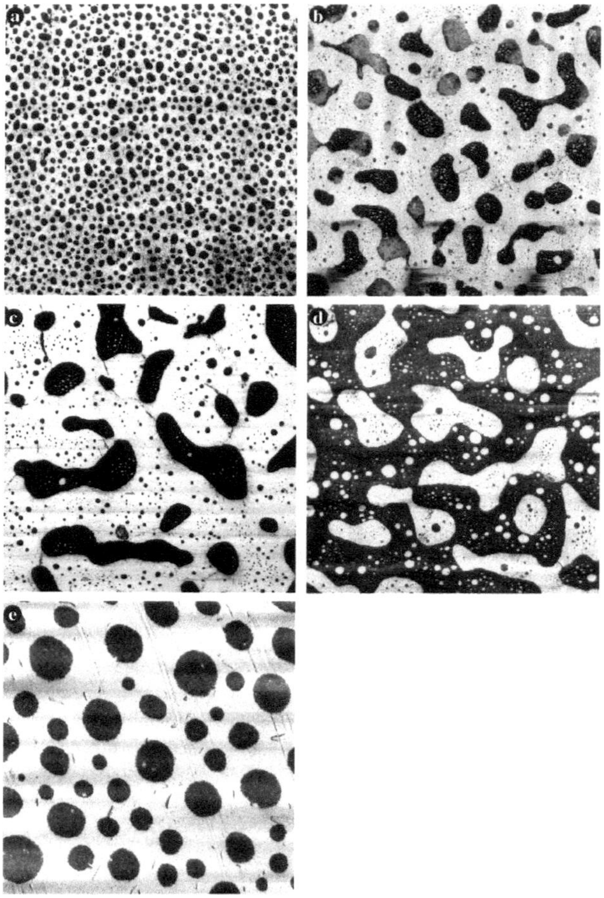

Figure 6.14 AFM phase images of i-PP/SEBS blends. The blend composition ratios of i-PP/SEBS are 80/20 wt.% in (a), 50/50 wt.% in (b) and (c), 30/70 wt.% in (d), and 50/50 wt.% in (e). The samples in (a) and (b) were prepared at 130 °C, the samples in (c) and (d) were annealed at 170 °C, and the sample in (e) was annealed at 190 °C and quenched to room temperature. The scan size was 16 μm in (a), 30 μm in (b)–(d), and 5 μm in (e). The contrast covers phase angle variations in the 100° range in all images. Reproduced with permission from [58]. Copyright 1998, American Chemical Society

■ 6.10 Conclusions

Phase-separation-induced various morphologies in binary polymer blends are significant for many applications. The key challenge here is that the morphologies generated during phase separation are not stable; hence, compatibilizers are in great demand. Many techniques have been developed to assess the thermal, mechanical, and electrical properties of phase-separated polymer blends. However, phase behavior detected by one or two techniques may not be sufficient to explain the detailed microstructure of the binary component. A single phase detected by DMA and DSC measurements does not always mean that the blends are necessary miscible when looked at from AFM. Hence a combination of techniques is required to explore the in-depth structural properties of the binary component to correlate with their final properties.

References

[1] Tan, Y., Guo, M. Study on the phase behavior of asphalt mastic. *Constr. Build. Mater.* 47, 311–317 (2013)

[2] Fox, T. G. Influence of diluent and of copolymer composition on the glass temperature of a polymer system. *Bull. Am. Phys. Soc.* 1, 123 (1956)

[3] Gordon, M., Taylor, J. S. Ideal copolymers and the second-order transitions of synthetic rubbers. I. Non-crystalline copolymers. *J. Appl. Chem.* 2, 493–500 (1952)

[4] Kwei, T. The effect of hydrogen bonding on the glass transition temperatures of polymer mixtures. *J. Polym. Sci., Polym. Lett. Ed.* 22, 307–313 (1984)

[5] Zuideveld, M. et al. Miscibility and properties of linear poly(L-lactide)/branched poly(L-lactide) copolyester blends. *Polymer* 47, 3740–3746 (2006)

[6] Na, Y.-H. et al. Compatibilization effect of poly(ε-caprolactone)-b-poly(ethylene glycol) block copolymers and phase morphology analysis in immiscible poly(lactide)/poly(ε-caprolactone) blends. *Biomacromolecules* 3, 1179–1186 (2002)

[7] Hiljanen-Vainio, M., Varpomaa, P., Seppälä, J., Törmälä, P. Modification of poly(L-lactides) by blending: mechanical and hydrolytic behavior. *Macromol. Chem. Phys.* 197, 1503–1523 (1996)

[8] Yang, J.-M., Chen, H.-L., You, J.-W., Hwang, J. C. Miscibility and crystallization of poly(L-lactide)/poly(ethylene glycol) and poly(L-lactide)/poly(ε-caprolactone) blends. *Polym. J.* 29, 657–662 (1997)

[9] Dell'Erba, R., Groeninckx, G., Maglio, G., Malinconico, M., Migliozzi, A. Immiscible polymer blends of semicrystalline biocompatible components: thermal properties and phase morphology analysis of PLLA/PCL blends. *Polymer* 42, 7831–7840 (2001)

[10] Sheth, M., Kumar, R. A., Davé, V., Gross, R. A., McCarthy, S. P. Biodegradable polymer blends of poly(lactic acid) and poly(ethylene glycol). *J. Appl. Polym. Sci.* 66, 1495–1505 (1997)

[11] Coleman, M. M., Painter, P. C., Graf, J. F. *Specific Interactions and the Miscibility of Polymer Blends.* CRC Press, Boca Raton (1991)

[12] Eguiburu, J., Iruin, J., Fernandez-Berridi, M., San Román, J. Blends of amorphous and crystalline polylactides with poly(methyl methacrylate) and poly(methyl acrylate): a miscibility study. *Polymer* 39, 6891–6897 (1998)

[13] Zhang, G., Zhang, J., Wang, S., Shen, D. Miscibility and phase structure of binary blends of polylactide and poly(methyl methacrylate). *J. Polym. Sci., Part B: Polym. Phys.* 41, 23–30 (2003)

[14] Abiad, M.G., Campanella, O.H., Carvajal, M.T. Assessment of thermal transitions by dynamic mechanical analysis (DMA) using a novel disposable powder holder. *Pharmaceutics* 2, 78–90 (2010)

[15] *Polymer Blends Handbook, Vol. 1.* Utracki, L.A. (Ed.) (2002), Kluwer, Dordrecht

[16] Schneider, H.A. Conformational entropy contributions to the glass temperature of blends of miscible polymers. *J. Res. Natl. Inst. Stand. Technol.* 102, 229 (1997)

[17] Martin, O., Avérous, L. Poly (lactic acid): plasticization and properties of biodegradable multiphase systems. *Polymer* 42, 6209–6219 (2001)

[18] Jacobsen, S., Fritz, H.-G. Plasticizing polylactide–the effect of different plasticizers on the mechanical properties. *Polym. Eng. Sci.* 39, 1303–1310 (1999)

[19] Wang, L., Ma, W., Gross, R., McCarthy, S. Reactive compatibilization of biodegradable blends of poly(lactic acid) and poly(ε-caprolactone). *Polym. Degrad. Stab.* 59, 161–168 (1998)

[20] Wang, R., Wang, S., Zhang, Y., Wan, C., Ma, P. Toughening modification of PLLA/PBS blends via in situ compatibilization. *Polym. Eng. Sci.* 49, 26–33 (2009)

[21] Park, J.W., Im, S.S. Phase behavior and morphology in blends of poly(L-lactic acid) and poly(butylene succinate). *J. Appl. Polym. Sci.* 86, 647–655 (2002)

[22] Li, Y., Shimizu, H. Toughening of polylactide by melt blending with a biodegradable poly(ether) urethane elastomer. *Macromol. Biosci.* 7, 921–928 (2007)

[23] Qiu, J. et al. Miscibility and double glass transition temperature depression of poly(L-lactic acid) (PLLA)/poly(oxymethylene)(POM) blends. *Macromolecules* 46, 5806–5814 (2013)

[24] Liu, T.-Y., Lin, W.-C., Yang, M.-C., Chen, S.-Y. Miscibility, thermal characterization and crystallization of poly(L-lactide) and poly(tetramethylene adipate-co-terephthalate) blend membranes. *Polymer* 46, 12586–12594 (2005)

[25] Nurkhamidah, S., Woo, E.M. Phase separation and lamellae assembly below UCST in poly(L-lactic acid)/poly(1,4-butylene adipate) blend induced by crystallization. *Macromolecules* 45, 3094–3103 (2012)

[26] Komalan, C., George, K.E., Kumar, P.A.S., Varughese, K.T., Thomas, S. Dynamic mechanical analysis of binary and ternary polymer blends based on nylon copolymer/EPDM rubber and EPM grafted maleic anhydride compatibilizer. *eXPRESS Polym. Lett.* 10, 641–653 (2007)

[27] Polios, I.S. et al. Late stages of phase separation in a binary polymer blend studied by rheology, optical and electron microscopy, and solid-state NMR. *Macromolecules* 30, 4470–4480 (1997)

[28] Brannock, G., Barlow, J., Paul, D. Blends of styrene/maleic anhydride copolymers with polymethacrylates. *J. Polym. Sci., Part B: Polym. Phys.* 29, 413–429 (1991)

[29] Chopra, D., Kontopoulou, M., Vlassopoulos, D., Hatzikiriakos, S.G. Effect of maleic anhydride content on the rheology and phase behavior of poly(styrene-co-maleic anhydride)/poly(methyl methacrylate) blends. *Rheol. Acta* 41, 10–24 (2002)

[30] Kapnistos, M. et al. Rheology of a lower critical solution temperature binary polymer blend in the homogeneous, phase-separated, and transitional regimes. *Macromolecules* 29, 7155–7163 (1996)

[31] Sasabe, H., Saito, S., Asahina, M., Kakutani, H. Dielectric relaxations in poly(vinylidene fluoride). *J. Polym. Sci., Part A-2: Polym. Phys.* 7, 1405–1414 (1969)

[32] Sengwa, R., Choudhary, S. Dielectric properties and fluctuating relaxation processes of poly(methyl methacrylate) based polymeric nanocomposite electrolytes. *J. Phys. Chem. Solids* 75, 765–774 (2014)

[33] Seki, Y., Kita, R., Shinyashiki, N., Yagihara, S., Yoneyama, M. In *AIP Conference Proceedings*. Vol. 1518, 466–469 (American Institute of Physics, 2013)

[34] Sy, J.W., Mijovic, J. Reorientational dynamics of poly(vinylidene fluoride)/poly(methyl methacrylate) blends by broad-band dielectric relaxation spectroscopy. *Macromolecules* 33, 933–946 (2000)

[35] Yang, J.-h. et al. Structural relaxation and dielectric response of PVDF/PMMA blend in the presence of graphene oxide. *Polymer* 229, 123998 (2021)

[36] Madbouly, S. A. Broadband dielectric spectroscopy for poly (methyl methacrylate)/poly (α-methyl styrene-co-acrylonitrile) blend. *Polym. J.* 34, 515–522 (2002)

[37] Zhang, S., Painter, P. C., Runt, J. Coupling of component segmental relaxations in a polymer blend containing intermolecular hydrogen bonds. *Macromolecules* 35, 9403–9413 (2002)

[38] Jin, X., Zhang, S., Runt, J. Broadband dielectric investigation of amorphous poly(methyl methacrylate)/poly(ethylene oxide) blends. *Macromolecules* 37, 8110–8115 (2004)

[39] Sengers, W., Van den Berg, O., Wübbenhorst, M., Gotsis, A., Picken, S. Dielectric spectroscopy using dielectric probes: a new approach to study glass transition dynamics in immiscible apolar polymer blends. *Polymer* 46, 6064–6074 (2005)

[40] Jiang, M., Chen, W., Yu, T. Controllable specific interactions and miscibility in polymer blends: 3. Non-radiative energy transfer fluorescence studies. *Polymer* 32, 984–989 (1991)

[41] Tiwari, R. R., Paul, D. Effect of organoclay on the morphology, phase stability and mechanical properties of polypropylene/polystyrene blends. *Polymer* 52, 1141–1154 (2011)

[42] Wang, Y., Xiao, Y., Zhang, Q., Gao, X.-L., Fu, Q. The morphology and mechanical properties of dynamic packing injection molded PP/PS blends. *Polymer* 44, 1469–1480 (2003)

[43] Lee, J. K., Han, C. D. Evolution of polymer blend morphology during compounding in an internal mixer. *Polymer* 40, 6277–6296 (1999)

[44] Galloway, J. A., Jeon, H. K., Bell, J. R., Macosko, C. W. Block copolymer compatibilization of cocontinuous polymer blends. *Polymer* 46, 183–191 (2005)

[45] Xavier, P., Jain, S., Chatterjee, K., Bose, S. Designer porous antibacterial membranes derived from thermally induced phase separation of PS/PVME blends decorated with an electrospun nanofiber scaffold. *RSC Adv.* 6, 10865–10872 (2016)

[46] Gebrekrstos, A., Sharma, M., Madras, G., Bose, S. New physical insights into shear history dependent polymorphism in poly(vinylidene fluoride). *Cryst. Growth Des.* 16, 2937–2944 (2016)

[47] Wang, B. et al. Critical composition of the β form of poly(vinylidene fluoride) in miscible crystalline/crystalline blends. *J. Phys. Chem. B* 119, 14303–14308 (2015)

[48] Mao, H.-j. et al. Constructing a microcapacitor network of carbon nanotubes in polymer blends via crystallization-induced phase separation toward high dielectric constant and low loss. *ACS Appl. Mater. Interfaces* 12, 26444–26454 (2020)

[49] Kajiyama, T. et al. Imaging of dynamic viscoelastic properties of a phase-separated polymer surface by forced oscillation atomic force microscopy. *Macromolecules* 27, 7932–7934 (1994)

[50] Paige, M. F. A comparison of atomic force microscope friction and phase imaging for the characterization of an immiscible polystyrene/poly(methyl methacrylate) blend film. *Polymer* 44, 6345–6352 (2003)

[51] Magonov, S. N., Reneker, D. H. Characterization of polymer surfaces with atomic force microscopy. *Annu. Rev. Mater. Sci.* 27, 175–222 (1997)

[52] Gedde, U. W. *Polymer Physics*. Springer, Dordrecht (1995)

[53] Wu, G., Miura, T., Asai, S., Sumita, M. Carbon black-loading induced phase fluctuations in PVDF/PMMA miscible blends: dynamic percolation measurements. *Polymer* 42, 3271–3279 (2001)

[54] Wen, G., Li, X., Liao, Y., An, L. Surface phase separations of PMMA/SAN blends investigated by atomic force microscopy. *Polymer* 44, 4035–4045 (2003)

[55] Dil, E. J., Carreau, P., Favis, B. D. Morphology, miscibility and continuity development in poly(lactic acid)/poly(butylene adipate-co-terephthalate) blends. *Polymer* 68, 202–212 (2015)

[56] Chen, C., Wang, J., Woodcock, S. E., Chen, Z. Surface morphology and molecular chemical structure of poly(*n*-butyl methacrylate)/polystyrene blend studied by atomic force microscopy (AFM) and sum frequency generation (SFG) vibrational spectroscopy. *Langmuir* 18, 1302–1309 (2002)

[57] Liu, T., Ozisik, R., Siegel, R.W. Phase separation and surface morphology of spin-coated films of polyetherimide/polycaprolactone immiscible polymer blends. *Thin Solid Films* 515, 2965–2973 (2007)

[58] Bar, G., Thomann, Y., Whangbo, M.-H. Characterization of the morphologies and nanostructures of blends of poly(styrene)-block-poly(ethene-co-but-1-ene)-block-poly(styrene) with isotactic and atactic polypropylenes by tapping-mode atomic force microscopy. *Langmuir* 14, 1219–1226 (1998)

[59] Cabral, J., Higgins, J., McLeish, T., Strausser, S., Magonov, S. Bulk spinodal decomposition studied by atomic force microscopy and light scattering. *Macromolecules* 34, 3748–3756 (2001)

[60] Prosycevas, I., Tamulevicius, S., Guobiene, A. The surface properties of PS/PMMA blends nanostructured polymeric layers. *Thin Solid Films* 453, 304–311 (2004)

7 Factors Influencing Phase Separation in Partially Miscible Polymer Blends

7.1 Introduction

Polymer blends are frequently classified as immiscible due to entropic factors that work against polymer blend miscibility. There are two forms of binary polymeric blends: blends with a co-continuous phase architecture and blends having a phase architecture as droplets scattered in a matrix. Co-continuous phase configuration is the cohabitation of two continuous structures within the same volume, each with its own internal network-like structure. Different continuities can be achieved by altering the mix compositions above particular threshold levels. Because of their unique phase structures, co-continuous polymer blends are suited to support a wide variety of purposes. Synergistic mechanical characteristics regulated electrical conductivity and selective permeability are all useful properties of co-continuous polymer blends.

Even though most polymeric materials are partially pairings, they are immiscible; polymer blends are nonetheless of great interest owing to their unique phase behavior. In partially miscible polymer blends, phase separation happens either by increasing the temperature, known as low critical solution temperature (LCST) behavior, or by reducing the temperature, known as upper critical solution temperature (UCST) behavior. The phase separation of LCST or UCST can occur through two separate processes: binodal decay/decomposition, which is thermodynamically metastable, as well as spinodal decay, which is thermodynamically unstable. The ultimate morphology of partially miscible blends is mostly governed by phase separation kinetics due to the complicated phase behavior [1].

Partially miscible polymeric blend phase separation is influenced by several factors such as solution mixing, doctor blade, as well as processing approaches like electrospinning, spin coating, compression molding, etc. The combined qualities of separate polymer components are not exactly reflected in the behavior of polymeric blends. The overall performance of the polymeric blends, as well as their properties, is determined by several factors: the individual properties of the polymer components (such as state, structure, and molecular weight), the preparation approach, handling temperature and time, polymer component conformation, morphology and structural considerations (interfacial area, shape, uniformity, size, as well as dispersal), along with the space between adjacently distributed units as well as bonding among polymeric entities.

The properties of the miscible polymeric blend(s) adhere to a numerical equation such as Equation 7.1:

$$X = X_{11} + X_{22} + I_{12} \tag{7.1}$$

Where X is the desired attribute, Φ denotes the concentration, while I denotes the interaction. When I is zero, polymeric blends might act as amalgams; if I is negative, it implies that antisynergistic effects in polymer blend properties can be detected; and if I is positive, synergistic effects in polymer blend properties can be noticed. For immiscible polymeric blends, the preceding expression is not applicable; their physical properties follow a different semiempirical law, which can be represented as Equation 7.2:

$$P_2 = \frac{1 + AB_2}{P_1 - B\psi_2} \tag{7.2}$$

Where P_1, denotes the continuous phase, P_2, denotes the dispersed phase, Φ_2 symbolizes the concentration of the dispersed phase, A the shape and orientation of the dispersed segment, B the relative values of the parameters P_1, P_2, and A along with ψ denotes packing ratio.

This chapter discusses the influence of these factors mentioned on phase-separated partially miscible polymeric blends.

■ 7.2 Processing Techniques

The main factors influencing the phase separation temperature in moderately miscible polymeric blends are discussed in this section.

7.2.1 Solution Mixing and Film Casting

The utilization of the solution mixing technique for the formulation of polymeric mixtures influences the ultimate attributes of the final material [2]. The impact of solvent–polymer interactions was explored, as this would otherwise reduce the prevalence of certain interactions in polymers within a given blend with strong polar groups [2]. In this approach, the polymeric matrix/matrices are dissolved in a standard solvent while being vigorously stirred, then polymeric mixture is cast by solvent removal either using solvent evaporation or using a nonsolvent. The fundamental disadvantage of this technology is that it is only appropriate for solvent-soluble polymers. The recovery of the solvent is also a concern for the widespread usage of this solution-blending approach [3, 4]. The architecture arising from solution mixing/film casting of a membrane is important to understand since it affects its effectiveness. It is widely understood that the morphological architecture is affected mostly by the rate of solvent outer diffusion into a trough containing water as well as the pace of water inner diffusion into the casting solution, delayed precipitation results in sponge-like pores, whilst quick precipitation results in finger-like pores (Figure 7.1). The phase equilibrium (thermodynamic characteristics), as well as diffusive characteristics of solvent plus nonsolvent, can be used to justify the demixing rate (kinetic properties). The phenomenon has been recently proven in PES@GO and PVDF@GO blend matrix films by Mohsenpour et al. [5]. Park et al. proposed that solvent casting has the ability to trap polymeric blends composed of binary diblock copolymer combinations in a metastable alternating gyroid phase [6]. They demonstrated, using self-consistent field theory (SCFT), that equimolar polymeric mixtures of double-gyroid-creating AB, as well as BC diblock copolymers having a comparatively immiscible A plus C blocks, should form an alternating gyroid architecture having complementary 3D A along with C architecture with nearly deviant free energy as well as dual phase-separated twofold gyroid states [6]. The inclusion of small proportion (1% of the total) of ABC triblock terpolymer creates an alternate gyroid firmness gap in the ensuing ternary phase graph, as per their hypothesis. The surfactant-like stabilizing influence of the triblock is rather indifferent to its specific composition as long as the B-block forms a relatively long link linking the A-rich combined with C-rich systems. When compared to existing approaches for producing an oscillating gyroid phase within block polymeric matrices, this mixing strategy offers considerable synthetic, as well as material, processing advantages [6].

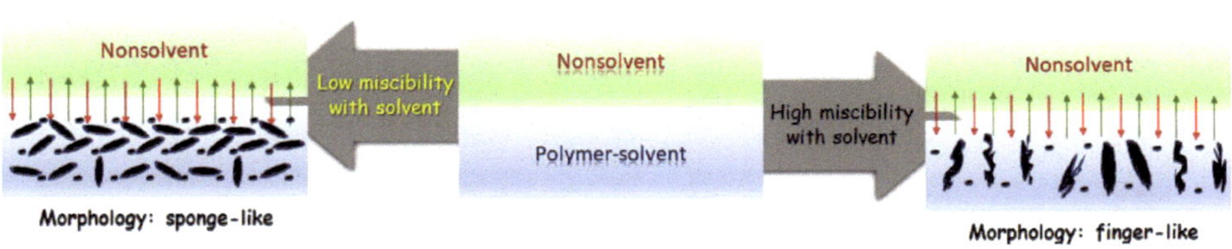

Figure 7.1 The influence of the solvent-nonsolvent interchange rate upon membrane morphology

Solution mixing has been reported by Tang et al. [3], where PVDF and PBA were liquified in DMF at an 80/20 mass ratio to make a solution having a wt.% of 20 mg/mL. The blended films were made by solution casting and then dried in a vacuum oven for 24 hours at 70 °C to thoroughly evaporate the solvent [3]. As a result of the solution mixing technique influenced phase separation in the blend, they showed from their study that the α phase of PVDF was transformed to its γ phase with the inclusion of PBA in situ during blending; the processing technique enhanced the transition of the α to γ phase from the center as it propagated along the radial axis of the main α-spherulite. The solution blending enabled the intermolecular interfacial interaction, which resulted in the α-to-γ phase transition, as well as the processing approach, coupled with its crystallization temperature, which led to the change in α lamina of PVDF to its γ lamina [3]. As PVDF has a T_m lower than that of the solid-to-solid transition, it represents a solid-to-solid transformation. However, they showed the phase transformation to be notable because the "α-to-γ" phase transition could result owing to changes in lamellar morphology, which contradicts the proposed solid-to-solid transition mechanism. Figure 7.2 depicts the phase transformation as according to the findings of Tang et al. [3].

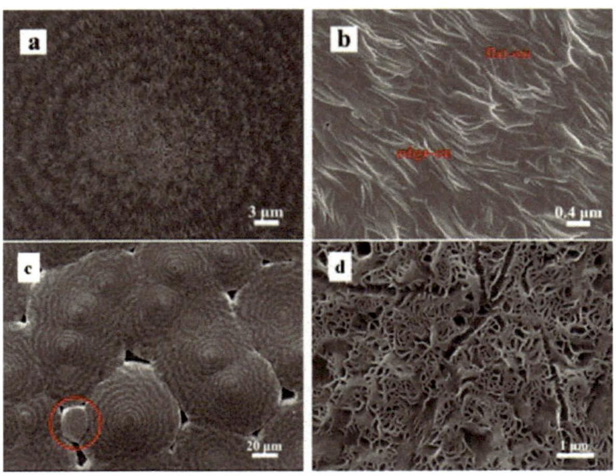

Figure 7.2 SEM micrographs of a banded α-PVDF spherulite (a) anc its magnifications (b). The neat PVDF was isothermally crystallized at 155 °C for 240 min and (c) SEM images of PVDF/PBA blend film and (d) the red circled part in (c). The sample was crystallized at 157 °C for 12 h. The red circle indicates a γ-PVDF spherulite. (For interpretation of the references to color in this figure legend, the reader is referred to the Web version of this article.) Reprinted with permission from Tang et al. [3]. Copyright 2022, Elsevier Science Ltd

Dong and Ozaki, in a study of PVPh and PMMA blends, concluded that even though the hydrogen bond between poly(4-vinylphenol) (PVPh) and PMMA favors their miscibility, phase separation still occurs due to the great difference in the two solvent–polymer interaction parameters, resulting in a non-equilibrium multiphase system that is effectively frozen [2]. They found that miscibility can be accomplished when ketones, such as methyl isobutyl ketone or methyl ethyl ketone, are employed as the preferred solvents; however, using tetrahydrofuran (THF) as the solvent gave an immiscible cast mix, as shown by NMR spectra [2]. Different miscibility is thought to be attributable to differences in the solvent-to-polymer interfacial characteristics in solution-cast blend systems ($\Delta\chi$). Poor miscibility is observed for a solvent having a high-level $\Delta\chi = |\chi_{12} - \chi_{13}|$, where χ_{12} and χ_{13} are the solvent-to-polymer interfacial parameters of the binary component homopolymers. As a result, it is best to use a solvent with approximately identical solvent-polymer interaction characteristics for both polymers.

Jin et al. [4] established a simple way to increase the TC (thermal conductivity) of epoxy-based composite materials using exceptionally low thermal conductive filler loading using reaction-induced phase separation (RIPS) through the solution mixing technique. Through a urea-functionalization method, the TC filler boron nitride nanosheets (BNNSs) were effectively processed by a mechanical-chemical reaction and then bound to amino groups (NH_2), potentially speeding the anhydride ring-opening reaction. Furthermore, the kinetic of RIPS was enhanced by a reduction in relaxation time "τ" caused by the grain growth of epoxy-rich phases. The loading of the BNNSs-NH_2 causes the NH_2 groups bonded to the edge of the BNNSs to enhance the storage modules of the nanocomposite as well as the glass transition temperatures [4]. The heat transmission channel might be effectively self-constructed owning to the unbroken PEI-rich phase as well as the selective positioning of BNNSs-NH_2 along the interfacial region of the polymeric matrices, which boosts the thermal control usage of the nanocomposite. Just with 1 wt.% BNNSs-NH_2 load capacity, the through-plane TC of nanocomposite increased by 83% (between 0.18 and 0.33 W m^{-1} K^{-1}). Filler thermal conductive efficiency (η) of BNNSs-NH_2 reaches 189 and 83 based on volume fraction and weight fraction, respectively [4]. The changes in the hybrid material morphology due to phase separation arising from the RIPS are presented in **Chapter 5** (Figure 5.4 and 5.5).

7.2.2 Doctor Blade

The magnitude of membrane casting, specifically doctor blade polymeric blend extrusion on a small-batch scale as well as slot die casting upon a roll-to-roll (R2R) on a large scale, has also been found to impact membrane morphological characteristics due to phase separation occurring within the membrane structural architecture [7, 8]. Immiscible/miscible polymer blend membrane design and construction at the laboratory scale is done with a doctor blade/casting knife at a constant speed on static material. The dope solution is poured on the substrate manually and is spread over the substrate without pre-determined flow rate and velocity by hand. In contrast, slot-diecasting on a R2R encompasses programmed procedural parameters like polymer dope flowrate and velocity, so the polymeric dope is exposed to shear. As a result, the preferred production method may lead to variation in membrane architecture.

Doctor blade casting has become a popular primary film casting approach for laboratory-based research [9], but this is not always the preferred approach for making up membranes on an industrial level as membrane morphology is hugely reliant on dope solution viscosity, making it inappropriate for continuous casting [10, 11].

The shape, size, applied pressure, etc., of the doctor blade during processing has a significant impact on the properties of partially miscible polymeric blends. Reports in this regard are either rare or limited, as only a few immiscible/miscible blend processes involve the use of a doctor blade. The thickness of the film, which controls several other characteristics of immiscible polymeric mixtures, is controllable with the aid of a doctor blade [1, 12].

Reports have shown that when the slot die technique is used for the preparation of polysulfone (PSF) ultrafiltration films, the membrane has patterned finger-like architecture within the whole of the assistive strata of the film; when the doctor blade technique is used, the membrane has sponge-like architecture as depicted in Figure 7.3 [13]. Additionally, as illustrated in Figure 7.3, the cross-section of the architecture of films varied depending on the production technique as well as the solvent utilized, as a result each of the six cross-sectional photographs in Figure 7.3 of films are unique. Both M1 and M4, doctor blade, and slot die, display comparable cross-sectional architectures exhibiting finger-like porosity near the active layer and uneven microvoids inside the supporting layer while using the traditional solvent DMAc. When PolarClean was employed, the films exhibited varied architectures depending on the casting procedure. When the slot die technology was used, the film displayed patterned finger-like architecture across the membrane supporting layer; when the doctor blade technique was used, the membrane displayed a sponge-like architecture [13]. Under different casting procedures, the solvent mixture of PolarClean and GVL (Table 7.1) created films exhibiting diverse morphologies. M3 exhibited a distinct structure, whilst using a slot die, with a dense active coating on the surface, gave a layer of needle-like microvoids as well as a sponge-like supporting layer. M6 revealed a sponge-like architecture in the supporting layer when using the doctor blade.

When considering the thermodynamic plus kinetic aspects of the demixing process, the demixing route of the doped solutions throughout the nonsolvent bath is the primary factor shaping the cross-section structure [14]. A rapid demixing phase, for example, can generate finger-like architectures, whereas a prolonged demixing procedure would generate sponge-like structures [14]. In cross-section morphologies of the slot die films, M1 has greater microvoids than M2 or M3, whereas M4 exhibits finger-like architectures, and M5, as well as M6, have sponge-like edifices. It could be because the PSF/DMAc dope mixture had low viscosity, resulting in an abrupt demixing reaction in the nonsolvent bathtub. Still, the PSF/PolarClean as well as PSF/50% PolarClean-50% GVL dope formulations, seemed to have a higher viscosity, resulting in a prolonged demixing reaction [13].

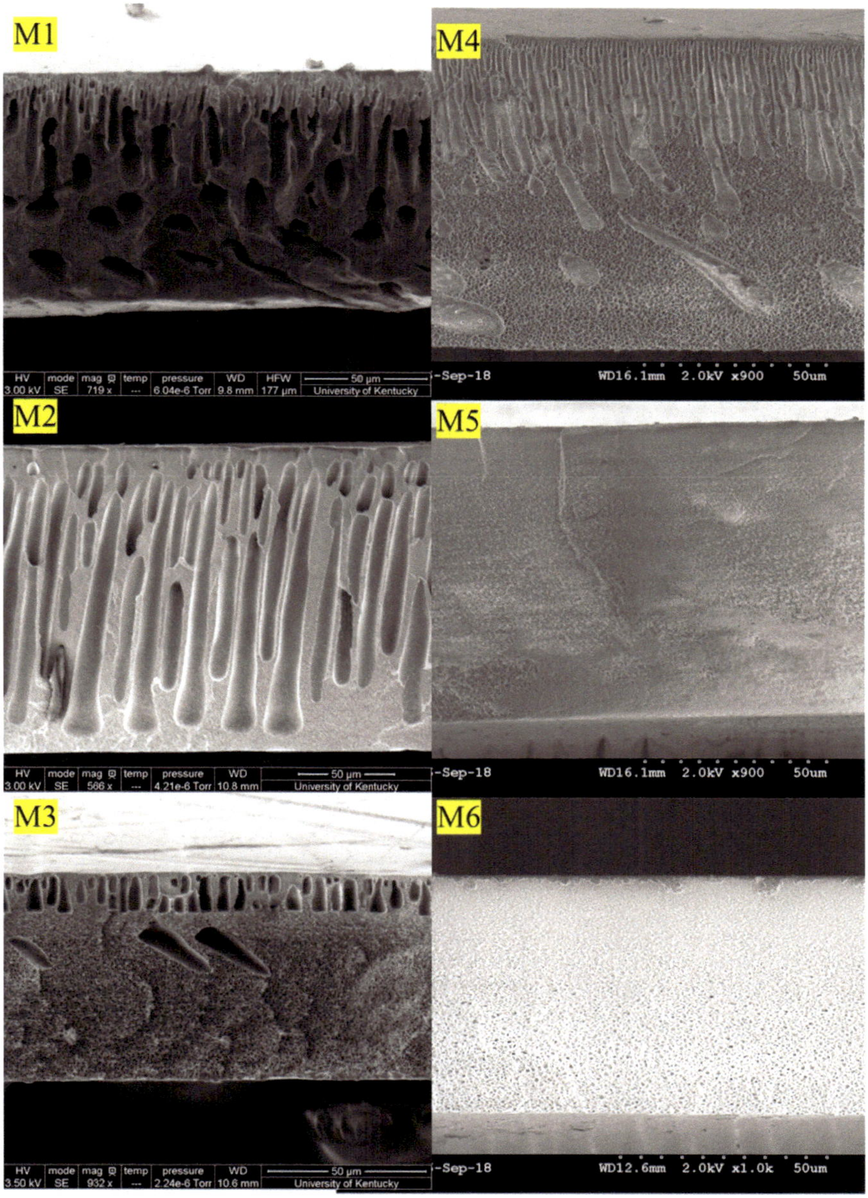

Figure 7.3 Cross-sectional SEM images of PSF membranes cast using different solvents and methods; for definitions of M1–M6 see Table 7.1. Reproduced with permission from Dong et al. [13]. Copyright 2020, Elsevier Science Ltd

Table 7.1 Formulations of PSF/Solvent Membranes[a] Reproduced with permission from Dong et al. [13]. Copyright 2020, Elsevier Science Ltd

Weight percent (%)	M1	M2	M3	M4	M5	M6
PSF	17	17	17	17	17	17
DMAc	83	–	–	83	–	–
PolarClean	–	83	50% × 83	–	83	50% × 83
GVL	–	–	50% × 83	–	–	50% × 83
	Slot Die Casting	Lab casting				

[a] All dope solutions are made using 17% PSF and 83% solvent.

Membranes produced by slot-die-coating are reported to be rougher when compared to those produced by the doctor blade technique of extrusion. This could be due to differences in these two manufacturing processes, as well as differences in system noise: as per the doctor blade technique, the blade is passed across the substrate top to create a liquid coating upon a rigid countertop, whereas slot-die casting is done upon a roll-to-roll approach. The liquid coating is propelled against a transiting substrate throughout the slot-die-coating process, which will be subjected to minor noise due to the motor drives [13]. The doctor-blade-produced sample presents a smoother finish due to cooperation of the doctor blade technique and the extremely fast de-mixing of the polymeric solution dope [13].

We believe that this phenomenon would repeat itself even in blends (immiscible/miscible) due to process-induced phase separation.

7.2.3 Electrospinning

Electrospinning is a simple electrohydrodynamic mechanism for producing fibers having diameters <100 nm, even as small as 5 nm [15]. An induced electric field causes a hanging droplet of polymer dope solution at the spin-pack to deform into a conical shape. In other words, the electrospinning process is capable of producing nanofibers using an electrically charged polymer solution as well as a polymer melts jet [16]. The first electrospinning setup made use of supporting electrodes for channeling the electrospinning jet towards the rotating take-up rollers/collectors. Researchers have also used various collection devices [17–19] and experimented with electric fields. In an interesting report, Dalton et al. [20], for example, formed electrospun poly(ε-caprolactone) (PCL) fibers between two collection rings in an inventive approach to create a stratified tissue architecture.

The type of electrospinning used in the preparation of electrospun polymeric blends impacts the character of the final product. For example, the use of horizontal electrospinning, vertical electrospinning, converse electrospinning, centrifugal electrospinning, near field electrospinning, needleless electrospinning, emulsion electrospinning, as well as changes in the spinneret resulting in coaxial electrospinning, co-electrospinning, in-line polymer blending, etc. all influence the properties of the immiscible polymeric blend to a large extent, as the processing techniques differ in every respect [1]. Approaches for fabricating novel fibers such as nonsolvent-induced phase separation (NIPS) and phase separation via electrospinning have been investigated [21, 22].

In the case of electrospun immiscible polymeric blends, the evaporation rate variance causes phase-separation within the blend. It has been demonstrated that H_2O vapor in the air can cause phase-separation, resulting in porosity within the fiber bulk, via vapor-induced phase-separation (VIPS) [23]. Nonsolvent-induced phase separation (NIPS), as well as thermally induced phase-separation (TIPS) techniques, have been widely utilized for the fabrication of porous membranes (thin film-based) [14, 24–26]. The phase-separation in NIPS, likewise referred to as immersion precipitation [14], is stimulated by solvent/nonsolvent exchange, whereas the phase-separation in TIPS is triggered by the cooling of the polymer. The polymer(s) precipitate swiftly upon dipping in a nonsolvent trough during NIPS, resulting in an extremely porous architecture [14]. Nevertheless, these phenomena have not been extensively investigated in conjunction with ES for the production of porous electrospun nanofibers. In an interesting example, McCann et al. [27] used a liquid nitrogen trough as the quenching intermediate to exploit the role of TIPS in the formation of highly porous electrospun nanofibers. TIPS were formed when the fibers were frozen in the liquid nitrogen trough, and afterwards the fibers were gently vacuum-dried, resulting in porous NFs [27].

It has been postulated that during the electrospinning process of core-shell fibers, the complex core of nanofibers is formed by impulsive phase separation [22]. Porous nanofibers with a specific topology can be created by using specific solvents, solvent blends, or polymeric blends in regulated ecological channels. For example, immiscible polymeric matrices in a generic solvent are then electrospun. In contrast, one of the polymer matrices is dissolved from the ES spun substrate to produce porous nanofibers: this system relies on phase separation affected by varying vaporization rates [22].

There are reports of differences in the properties of electrospun polymeric blends or their composites arising from differences in the processing techniques when electrospinning was compared with other polymer blend processing techniques [18]. It is interesting to note that core-sheath NFs could be fabricated via conventional ES established upon one spin pack by induced phase-separation between binary matrices in a uniform solution or an emulsion [28–30]. In the case of a unified solution, phase-separation is typically caused by the difference in solubility of the two polymeric matrices within the common solvent in the course of jet drying [31]. Because PEO and chitosan have different solubilities in water,

PEO@chitosan core-sheath nanofibers have been fabricated via electrospinning their uniformly mixed solution in H_2O (l) [32]. In order to trigger mismatch within the binary polymeric matrices, mixed solvents can be used. In a particular case, PCL@poly(*N*-isopropylacrylamide) (PNIPAAm) core-sheath NFs were spun by ES from a PCL and PNIPAAm solution in DMF/chloroform [33]. Because DMF has a higher boiling point (153 °C) than chloroform (61 °C), a higher proportion of DMF to $CHCl_3$ was created within the jet in conjunction with favored evaporation of $CHCl_3$. Given the variance in solubility of PNIPAAm in comparison to PCL in DMF, thermodynamically, phase-separation was preferred. The comparatively short molecular chains led to higher movement of PNIPAAm toward the surface of the jet, resulting in the formation of core-sheath architecture. Furthermore, by changing the solubility of the polymeric component in the jet, water vapor can cause phase-separation. Electrospinning a homogeneous solution of PAN and PVP in DMF under high relative humidity, for example, resulted in the formation of PAN@PVP core-sheath NFs [34]. However, for most instances, the motions of the two polymeric matrix molecules within the jet are meaningfully reduced by the rapid vaporization of the solvent(s), hindering the substantial-scale phase-separation required for the formation of core-sheath NFs.

In the case of emulsion ES, whether oil-in-water as well as water-in-oil, phase-separation can be induced within the jet via de-emulsification triggered by solvent evaporation, leading to the creation of core-sheath NFs. A water-in-oil emulsion was created by combining the aqueous form of PEO in H_2O as well as the oil-phase of the amphiphilic PEG@PLA diblock copolymer in solvent (chloroform); ES gave PEO@PEG–PLA core-sheath nanofibers [35]. In some cases, the emulsion can be created by combining two polymers in a single solvent to create core-sheath NFs. After storing for one day, a blend of PMMA and PAN in DMF disintegrated into PMMA globules scattered in a PAN polymer solution [36]. It is also possible to disperse a large number of cross-linked polymers as enlarged colloids in a polymeric matrix solution for the preparation of an emulsion for the formation of core-sheath NFs. The phase-separation between the colloids as well as the condensed polymeric matrices/matrix results in the creation of core-sheath NFs [37]. In one typical example, PNIPAAm microgel colloids formed an arrangement in the central area of PCL fibers [37]. The main disadvantages of emulsion electrospinning are the instability of the emulsified solution and the presence of an emulsifier, which may have an impact on the applications of the nanofibers [38]. Furthermore, emulsion electrospinning can produce NFs in which the dispersed portion is implanted as separate globules rather than as a core-sheath configuration.

The ES process is a viable means for the preparation of immiscible/miscible phase-separated polymeric blends.

7.2.4 Spin Coating

Spin-coating is a popular method for producing even thin polymeric membranes. For example, the functional sheets in most investigational semiconducting polymeric-based gadgets, such as LEDs and photovoltaics, have been made in this manner. The effectiveness of such gadgets could be increased by utilizing polymer mixes; these undergo phase separation during the spin-coating procedure, resulting in complex morphology that leads to increased function [39].

Previously, it was hypothesized that phase separation in immiscible polymeric blends could be traced back to the processes observed in bulk. In 2005, Heriot and Jones showed that the process is very intricate, involving the development of a stratified structure and then potentially splitting up into adjacent domains using spin coating [39].

Spin-coated thin films of polymer mixtures could phase separate in two ways: by creating adjacent domains or by splitting into separate layers. The later situation (self-stratification/vertical phase-separation) can be beneficial in a variety of practical uses, including polymer-based optoelectronics.

Spin coating (also spin-coated polymer blend samples) is affected by several factors like capillary effects, solution radial flow, as well as an unsteady air-flow field over the polymeric membrane. These factors, to a great extent, affect the phase separation of polymeric blends prepared using this technique [40].

Polymers are intrinsically immiscible in most cases, such as spin coating, and phase-separation occurs as soon as polymeric blends are removed from high temperatures or solution. In general, immiscible polymeric blends have segregated architecture with areas scattered within a continuous phase [41]. The phase-separation phenomena in polymeric blends with regard to thin films may differ significantly from that of bulk polymer blends. In thin membranes/films, the two faces (air/film and film/substrate) could have a significant impact on phase separation. The chemical nature, as well as surface energy of the material, also play a vital part in phase separation. On various substrates, diverse surface structural morphologies have been examined, along with well-patterned surfaces; architectures have also been acquired via controlling the material surface [42–44].

Heriot and Jones [39] showed in their study of immiscible phase-separated spun cast thin films that interfacial volatility in a transitory wetting layer resulting in lateral phase-separation. Their findings show that a binary polymeric mixture first undertakes vertical stratification, after which the boundary between the layered sheets is unstable, resulting in the final phase-separated thin film: it has provided the foundation for the development a complete mechanistic understanding of the creation of the desired morphology in thin blended polymer membranes, leading to the balanced design of fabricating conditions to achieve desirable morphologies via self-assembly [39].

Liu et al. [40] investigated the surface morphology in thin films of immiscible polyetherimide-PCL blends utilizing SEM along with AFM. Spin-coating from dichloromethane solution was utilized by the authors to produce the thin films. During the early stages of spin-coating, a self-built periodic array of phase-separated areas was detected, which was caused by capillary influences, solution radial-flow, as well as an unsteady air-flow field over the membrane. A secondary phase separation was observed during the solvent evaporation stage of spin-coating. The differences in surface topographies of the two distinct phases are attributed to different solvent evaporation rates within each phase [40]. A wide range of domain structural architecture as well as surface morphologies were detected as a function of the polymer blend.

Ebbens et al. [45] showed that spinodal decomposition dominates the phase-separation phenomenon in PS/PI (1:1) blends, leading to the appearance of a bicontinuous structure around 4 s before the membrane was entirely dried. As the film dried, the surface motif roughens, but there was slightly more lateral development at the studied length-scale of 0 to 250 nm. Their findings agree with the theory established by Buxton et al. [46], predicting that initial phase separation will occur at the upper surface of the nascent membrane wherein the concentration of the solvent is lowest, and then the creation of the pattern as the inherent material phase-separates. They also gave evidence of a secondary phase separation process in the closing fraction of a second (see Figure 7.4), which agrees with the limit diffusion placed upon the last solvent vaporization. Furthermore, Marangoni instabilities were assigned a higher scale surface irregularity earlier in the spin-coating technique but were said to have a negligible impact on the final morphological structure as well as phase architecture in their system [45].

Guo et al. reported a new lithography strategy with respect to phase-separation of PS/PEG immiscible polymer blends via spin coating [47]. They demonstrated that the morphological guiding factors of the porous nanostructure resulting from varying phase separation could be effectively tuned by regulating the weight ratio of the PS/PEG polymeric blends, their concentration, and the spin-coating speed [47]. They also showed that the spin-coating velocity could influence the structure of nanopores caused by the phase separation of PS/PEG mixtures in toluene. The SEM micrographs of phase-separated PS/PEG with diverse spin velocities are shown in Figure 7.5(a) to (c) and Figure 7.6 for a consistent PS/PEG weight ratio of 1:2 and a fixed concentration of 5 wt.% [47].

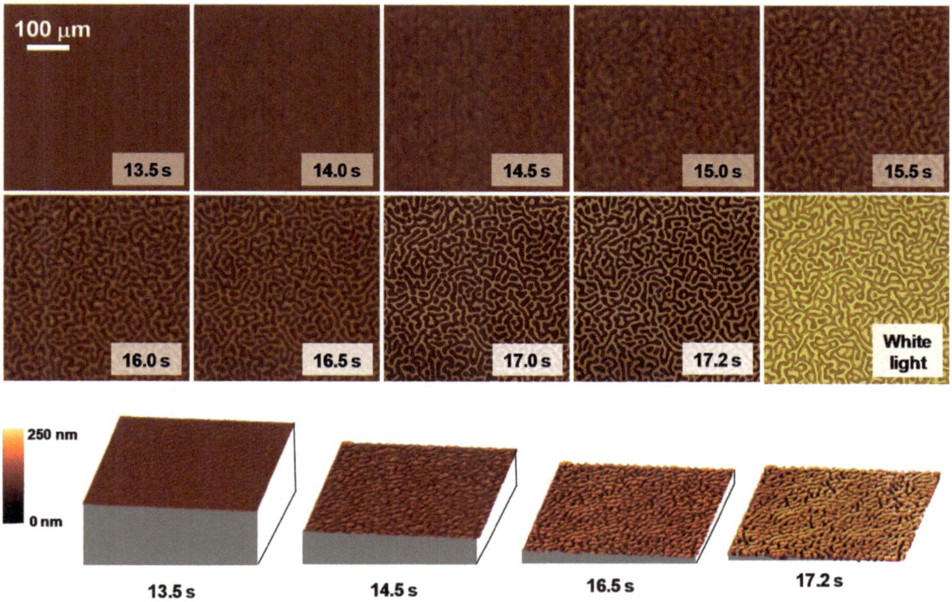

Figure 7.4 (Top) 2D relative height reconstructions during spin coating a PS/PI blend at 1500 rpm obtained from pixel by pixel drying curve arrays generated from fringe counting the stroboscopic optical images. (Bottom) 3D surface morphology reconstructions. Reproduced with permission from [45]. Copyright 2011, American Chemical Society

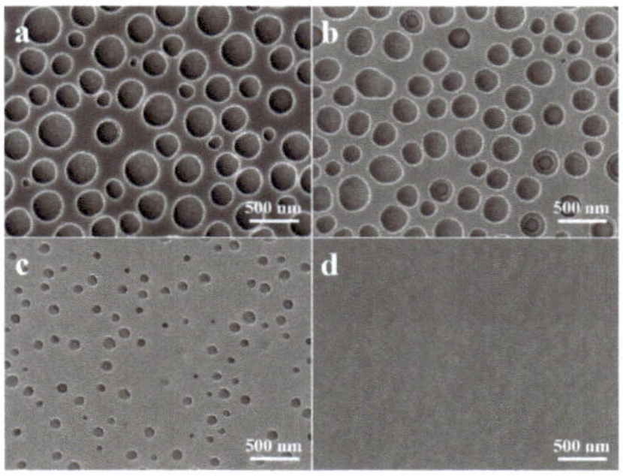

Figure 7.5 Top view of SEM images of PS porous film with different polymer blend concentrations. (a) 5 wt.%; (b) 4 wt.%; (c) 3 wt.%, and (d) 2 wt.%. The initial PS/PEG weight ratio was set at 2:3, and the spin speed was 3000 r·min^{-1}. Reproduced with permission from [47]. Copyright 2011, American Chemical Society

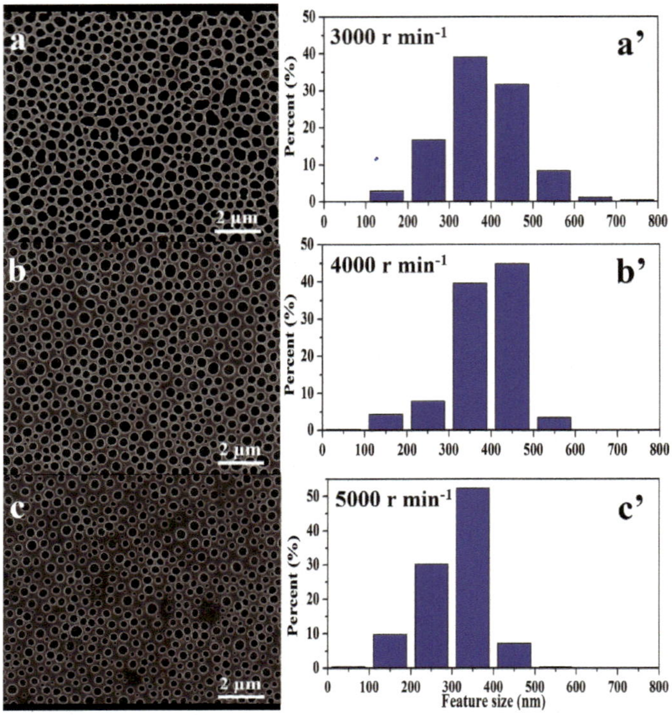

Figure 7.6 Top view of SEM images of PS nanopore structure produced with (a) 3000 r min^{-1}, (b) 4000 r min^{-1}, and (c) 5000 r min^{-1}. (a'), (b'), and (c') are the corresponding feature size distribution of PS nanopore structure. The PS/PEG weight ratio and concentration are fixed at 1:2 and 5%, respectively. Reproduced with permission from [47]. Coɔyright 2011, American Chemical Society

Mokarian-Tabari et al. [48] showed that by regulating the rate of evaporation through the course of the spin-coating procedure, they could achieve either self-stratification or phase-separation lateral within the same formulation(s). They related it to a previously proposed phenomenon for phase-separation through spin-coating in thin films, wherein a momentary wetting layer disintegrates due to a Marangoni-type unsteadiness triggered by a solvent concentration gradient within the desiccating films. Their findings showed that show that rapid evaporation leads to a laterally phase-separated structure, while reducing the evaporation rate suppresses the interfacial instability and leads to a self-stratified final film.

Karim et al. used spin coating to investigate the phase-separation of ultrathin polymer blend films of PS and poly(butadiene) on microcontact, printed alkanethiol patterns with hydrophobic and hydrophilic end groups (CH_3 and COOH) (Figure 7.7). Their simulations propose that the morphological phase-separation could be governed by forms that regulate the polymer-to-surface interaction, and this was supported experimentally [49]. The simulations offer insightful information into "pattern-directed spinodal decomposition",which aided their design of their statistics using a Cahn–Hilliard–Cook (CHC) conceptual framework of phase separation of tunable interfacial interactions on the outside of the substrate.

Figure 7.7 shows cross sections showing a simulation of composition variations in phase-separating blend films on a patterned substrate. Simulations employ CHC dynamics with a modulated boundary interaction defined in:

$$\frac{F_s}{k_B T} = \iint dxdy \left[h_0 \sin\left(\frac{2\pi x}{l_p}\right) \phi(x,y) + \left(\frac{1}{2}\right) g\phi^2(x,y) \right]$$

The surface pattern wavelength l_p is about twice the maximally unstable phase separation scale ~"spinodal wavelength") λ_0 in these simulations ($l_p \cong 2.2\ \lambda_0$). Variations in film height due to unequal surface and interfacial tensions, the deformability of the polymer-air boundary, and hydrodynamic interactions are not treated in this simulation.

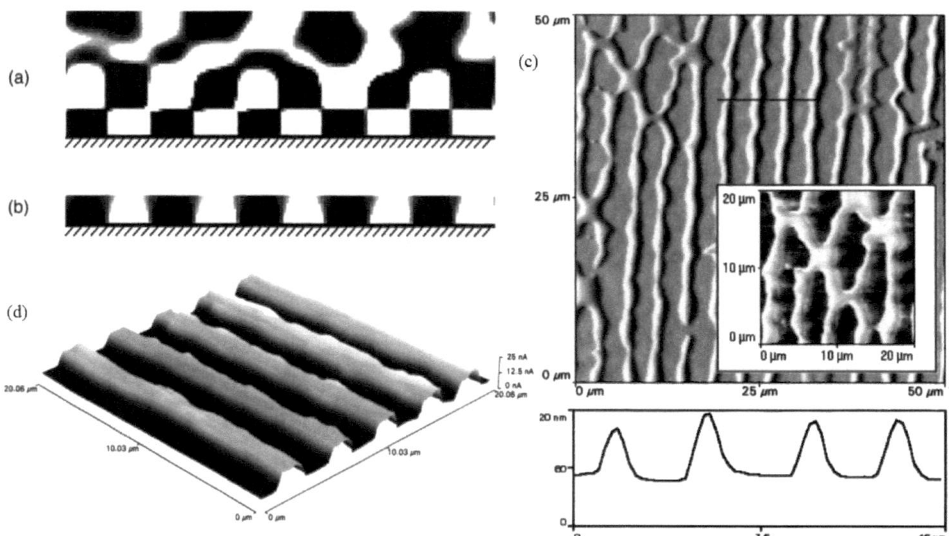

Figure 7.7 Cross sections showing a simulation of composition variations in phase-separating blend films on a patterned substrate. (a) "Checkerboard" pattern in thicker film cross sections arising from surface-directed phase separation in combination with pattern-directed lateral phase separation. (b) Thinner films exhibit only pattern-directed lateral phase separation. (c) Intermediate stage of dPS/PB blend film phase separation on a µCP patterned surface. AFM image of topography with accompanying line profile (1.5 h at 25 °C). Solid line on figure denotes line profile position. Inset shows lateral force image; lighter PB-rich regions have higher friction corresponding to the topography at the upper left-hand corner. (d) Lateral force AFM image in profile view late stage of dPS-PB blend film phase separation on a µCP patterned surface. Elevated regions correspond to PB-rich domains. The pattern is essentially unchanged after 24 h. Reproduced with permission from [49]. Copyright 1998, American Institute of Physics

7.2.5 Compression Molding

Processes including liquid drop elongating into threads, thread breakdown to tinier droplets, as well as droplet coalescence into bigger globules occur through immiscible polymeric matrices melt mixing, such as compression molding, resulting in the droplets and matrix morphology. The equilibrium of these rival processes affects the ultimate particle sizes of the polymeric blends, which impacts the qualities of the blend. Furthermore, characteristics such as the viscosity percentage, composition of the blend, elasticity percentage, shear stress, as well as interfacial tension are connected to droplet and matrix morphology, and droplet sizes [1].

In a study by Zhang et al. [50], a 50/50 wt.% PLA/PS blend was pressure shaped and toughened between dual equal plates without being restricted circumferentially in the course of the compression molding procedure. Different conditions for geometrical confinement, including varied gap size and compression ratio (initial to final thickness ratio), as well as modified surface properties, were utilized. They showed that the gap size played a profound role in affecting the structural development; the phase size near the mold surface was smaller than away from the surface. The actual phase structure and the resulting gradient in pore size were further affected by the mold surface properties. The actual phase structure and the resulting gradient in pore size were further affected by the mold surface properties.

Moreover, the pressure proportion was found to influence the morphological improvement, particularly close to the mold surface. At a very high molding pressure proportion, a slender sheet of PLA was shaped promptly in the course of applied pressure. The thickness of the layers was either enhanced or decreased in size, depending on the surface properties of the mold. Understanding of these geometrical constraining impacts and implementing them during processing immiscible polymeric blends may prompt the development of new materials [50].

Blends produced by compression molding result in a nonuniform phase separation/structure in immiscible polymeric blends. In this respect, Fortelný et al. [51] showed that compression-molded specimens of a variety of binary as well as ternary polymeric blends, blended extensively for an extended period in the Plasticorder chamber, resulted in a nonuniform structured phase made of regions with significantly diverse mean globule size [51]. They also found nonuniform structured phases in formulations that were immediately quenched after blending within the Plasticorder chamber, as well as in specimens blended in the twin-screw extruder [51].

Frone et al. examined the compression molding, 3D printing as well as extrusion study of morpho-structural, thermal, and mechanical characteristics of PLA-PHB-cellulose nanocrystals blend nanocomposites [52]. They established that the thermomechanical characterization of uncrosslinked, as well as crosslinked PLA-PHB blends as well as nanocomposites revealed the impact of the processing method. When compared to compression molded films, filaments acquired via extrusion, as well as 3D printed meshes had higher storage modulus values [52]. Similarly, TGA analysis revealed an enhancement in the onset degradation temperature of PLA-PHB mixtures and composites after extrusion and 3D printing by more than 10 °C when compared to compression molding. This is a clear indication that by selecting the correct processing method, PLA-PHB items with improved boundary adhesion, thermal stability, as well as mechanical character could be attained [52].

We also know that immiscible polymeric blends having a co-continuous structured phase could indeed produce material characteristics that are synergistic. Zheng et al. [53] investigated a 50/50 wt.% PLA-PS mixture with CPS as a model scheme to investigate the effect of compression molding conditions (consisting of compression rate, compression ratio, and mold surface properties) on the final structured phase of the molded part. They observed that whilst a high-level compression percentage, as well as a fast compression rate, were used during molding, the characterization at the microscopic level clearly established the rapid development of a PLA-rich sheet at the mold contact surface.

Bose et al. used compression molding as a vital tool in the processing of poly[(α-methylstyrene)-co-acrylonitrile]/poly(methyl methacrylate) (PαMSAN/PMMA) mixtures, with the assumption that the squash deformation flow in the course of compression molding allows the phase-separated regions that were unmixed during mixing/extrusion to once again blend, or at least convert into micro-heterogeneous phases [54]. They measured the T_g of the compression molded blends to get a measurable proof of this occurrence. All of the blends investigated by them had a single T_g, confirming the uniformity of the blends.

Choosing the appropriate processing method for immiscible polymeric blends can to a large extent, affect the phase separation and consequently the structure of these systems.

7.2.6 Incorporation of Nanoparticles

The primary information gained on nanoparticle-induced morphology transformation in multiphase systems, such as polymeric blends, has resulted in the formation of highly technological performance targets, including the fabrication of heat-resistant mixtures based on virgin or recycled plastics as well as biocompatible and biodegradable polymeric materials with prescribed topography as well as texture. A fivefold increase in publications on Scopus shows the growing interest in multiphase matrix composites compared to single-phased composites. There is significant scientific interest in the utilization of nanoscale particles to alter the structure of nonpolymeric\hybrid systems; significant instances of this kind of "soft matter" structure are low-molecular-weight emulsions, suspensions, and foams [55]. There is also much research is aimed utilizing immiscible phases as a liquid model for attempting to control the spatial structure of nanoparticles, ultimately realizing innovative nanoscale structures, in addition to understanding how to use nanofillers to steer the fabric morphological characteristics. Because these two mechanisms are so closely linked, understanding how particle inclusion affects the behavior of the blends in both solution and solid phases and, in turn, how the architectural evolution of the latter affects the spatial configuration of the filler, is an intricate area in the development of customized materials.

The necessity for such basic knowledge is strongly felt within the polymer blends area: although nanoparticle-influenced structural variations in such systems have also been widely discussed, most proposed hypotheses for the primary physical principles are contradictory or, at best, applicable only to the particular model under consideration.

The usefulness of fine materials as structural stabilizers in a variety of multiphase systems, such as polymeric blends, is well acknowledged. Pickering emulsions, which are emulsions with low viscosity with interfacially adsorbed colloidal particles, are a common example. Furthermore, when one of the media is air, the architectural constraint is imposed by the results of the filler in particle-stabilized foams as well as liquid drops such as "liquid marbles as well as dry liquids" [56]. In the case of immiscible polymeric blends, a common method for stabilizing both dispersed as well as co-continuous structures is the insertion (or production) of a compatibilizer. This is also a block copolymer that chelates at the polymer interface, or rather serves two purposes: (i) decreasing interfacial tension and so increasing droplet breakage during mixing (similar to emulsifiers in emulsions with low viscosity), and (ii) inhibiting future coalescence events [57]. Intensive research has been conducted over the last 20 years, with the most relevant results discussed here, has revealed that the selective nanoparticles localization within a polymeric mixture not only enhances the compatibility of polymeric phases, but it also causes a variety of morphological changes in diverse length scales.

Regrettably, synthesizing the critical compatibilizer for several polymer blends is challenging, if not unattainable, plus extra challenges arise when working with systems including more than two polymeric phases. Submicron nanoparticles have been shown to have a comparable but nonspecific compatibilizing effect, extending their utility far beyond that of simple reinforcing substances.

Composite materials can be made up of a variety of components, including not only various polymers but also plasticizers, stabilizers, and organic/inorganic particles.

Since 2010, the use of minute amounts of nanometer-sized particles has been found to provide performance benefits far greater than those provided by ordinary micrometer-sized particles. This is due to the enormous specific surface area of the nanoparticles, which can significantly alter processability, modulus, impact strength, physical ageing, and conductivity. Furthermore, nanoparticles have made important advances in the processing of polymeric blends: they could be utilized as a compatibilizer to stop domain coarsening as well as stabilize developing structures. However, more research is needed to ascertain the impact of nanoparticles on the overall phase behavior and performance of polymeric blends. In most instances, the nanoparticles, such as carbon nanomaterials, selectively localize themselves on the structure of the polymeric blend.

In 1995, Gubbels et al. identified the compatibilizing influence of nanoparticles within multiphase polymeric systems/blends in a PE/PS mixture containing trace amounts (5 wt.%) of carbon black [58]. The existence of compatibilizer particulate within polymeric blends reduced the size of the dispersed polymeric matrix phase; this has been demonstrated for diverse nanofillers adsorbed interfacially, including silica [59, 60], titanium dioxide [61], and layered silicates [62]; the latter can cause significant morphological changes. In organoclay-filled PC/PMMA systems, the T_g of the polymer components collapsed into a single T_g, which was attributed to the tiny size of the polymeric phases as well as their immobilization by robust interactions with the reinforcing filler [63]. Interfacial rheology has been found to be important in coalescence inhibition: nanoparticles encasing polymers or polymer boundaries could create robust viscoelastic coatings that act as boundary mobility transformers instead of traditional compatibilizers [64]. The primary processes anticipated to describe the compatibilizing function of nanoscale fillers in polymeric blends are as follows [65]: (i) reduction of interfacial tension, (ii) coalescence inhibition due to the existence of a solid barrier around the minor polymeric phase(s), (iii) change in viscosity percentage as a result of the existence of the nanofiller in one of the polymeric phases, (iv) the immobilization of the dispersed drops (or of the matrix) by the creation of a physical network of particles; and (v) adsorption interface of polymer molecules at the surface of the particle. The compounding, breakdown, and coalescence processes all compete, with the latter governed by unfavorable interactions among immiscible component molecular segments. The presence of unevenly distributed particles alters the interfacial environment, shifting the breakup/coalescence equilibrium in favor of the former. It is important to recognize that if the reinforcing filler favorably locates itself within the interfacial area between the dispersed drops then it can toughen the interfacial area, decelerate coalescence [66]. Huitric et al. [67] focused on clay-reinforced LDPE/PA12 blends and attributed the structural enhancements observed at various components to the diverse reinforcement dispersions and to the various mechanisms discussed for nanoscale fillers discussed above. While LDPE was used as the matrix, almost all of the filler is concentrated at the polymer-to-polymer boundary, and also domain size decrease is solely due to coalescence inhibition caused by the solid impediment effect. Also, while PA12 was the matrix, the filler was present in both the polymeric matrix and the interface, resulting in a dual effect: on the other hand, the nanoparticles in the matrix enhanced the viscosity as well as caused droplet breakup, the interfacially adsorbed platelets also inhibit coalescence. Similarly, Hong et al. [68] reported a fine-tuning of the structural architecture of nanoclay-reinforced blends of PE@poly(butylene terephthalate) (PBT). At low filler loading, nanoparticles congregate in the droplet interface, suppressing droplet amalgamation; at elevated filler concentrations, nanoparticles localize in a specific component depending on their affinity with the component, in this case PBT, altering the viscosity percentage and impacting droplet breakdown. Using rheological assessments, Elias et al. [69] discovered that the equalization process of silica-reinforced PP-PS mixtures led to a decrease of effective boundary strain in hydrophilic nanoparticles as well as the creation of a rigid layer preventing coalescence in the instance of hydrophilic nanoparticles [69]. The addition of nanoparticles to polymer blends, like traditional compatibilizers, has an asymmetric effect [70, 71]. Similarly, a specific nanoparticle may be capable of inhibiting coalescence in an A in B globule/matrix mixture, although not in a B in A mixture. Vermant et al. [59] found that SiO_2 nanoparticles decreased coalescence within a PIB-PDMS mixture only if PIB creates the dispersion phase, wherein it is attributed to PDMS favored hydration of the fillers. Following trials with a broader range of filler particles showed a far more pronounced asymmetry: depending on the continuous phase, the size of a drop could be increased or decreased depending on the particle type (i. e., either prevent or enhance coalescence) [72]. It has also been envisaged that asymmetry is well connected to the flow parameters encountered via the systems. Similarly, clay nanoparticles at the PS-PMMA mix boundary have been reported to cause morphological improvement during vigorous flows and coarsening during slow and extended flows due to the clustering of encapsulated PMMA molecules [73]. Aside from the broad ranges of coalescence advancement as well as suppression, solid reinforcing fillers may encourage a transitional phenomenon known as “arrested coalescence”, in which coalescing drops remain frozen, though also coalescing, as a result of the combination of nanoparticles that are either interfacially immobilized or restricted within the minor phase. Arrested coalescence is caused by a balance between the decrease in interfacial energy and the increase in elasticity that occurs during coalescence [74]. Another outstanding result of nanoparticle inclusion in polymeric blends, composites, or hybrid systems is the stability of nonspherical fluid regions. A large number of interfacially adsorbed spherical particles can obstruct the interface of nonpolymer systems, inhibiting or even stopping shape relaxation [63, 74]. If particles are preferentially located in any of the other fluid phases, the modulation of randomly oriented domains is also revealed [74]: the increased elasticity and viscosity of the host phase are attributed to the structural instability and breakup. When particles adhere nearly simultaneously at two liquid-liquid boundaries, they can connect across drops or bind them together to form clusters [72]. Bridging occurs when the continuous phase selectively wets nanoparticles, and it has significant rheological consequences, providing the mix with a solid-like characteristic [75].

The wetting coefficient (ω_α), which Sumita et al. [76] initially incorporated from Young's model, has become the most universally accepted principle for assessing the thermodynamically favored localization of fillers (Equation 7.3):

$$\omega_\alpha = \frac{\gamma_{\text{filler-A}} - \gamma_{\text{filler-B}}}{\gamma_{\text{A-B}}} \tag{7.3}$$

Where $\gamma_{\text{filler-A}}$ is the interfacial tension between the nanofiller and polymer A, the symbol filler $\gamma_{\text{filler-B}}$ is the interfacial tension between the nanofiller and polymer B, while symbol $\gamma_{\text{A-B}}$ stands for the interfacial tension between the two polymer parts in Young's equation. To forecast the thermodynamic propensity of nanofillers to localize in a given component/matrix of immiscible polymeric blends, the wetting coefficient (ω_α) can be determined. If $\omega_\alpha > 1$ the nanofillers are projected to be in polymer B; if $\omega_\alpha < 1$, the nanofillers are projected to be in polymer A; whereas if $-1 < \omega_\alpha < 1$, the nanofillers are predicted to be at the interface of the blend. The harmonic-mean expression or the geometric-mean expression can be utilized to compute the relevant interfacial tensions. Equation 7.4 determines the interfacial tension $\gamma_{1\text{-}2}$ between component 1 and component 2 for the harmonic-mean equation:

$$\gamma_{1-2} = \gamma_1 + \gamma_2 - \frac{4\gamma_1^d \gamma_2^d}{\gamma_1^d + \gamma_2^d} - \frac{4\gamma_1^p \gamma_2^p}{\gamma_1^p + \gamma_2^p} \tag{7.4}$$

Equation 7.5 calculates the interfacial tension $\gamma_{1\text{-}2}$ for the Geometric-mean equation:

$$\gamma_{1-2} = \gamma_1 + \gamma_2 - 2\left(\sqrt{\gamma_1^d \gamma_2^d} + \sqrt{\gamma_1^p \gamma_2^p}\right) \tag{7.5}$$

Where γ_1, γ_2 denotes the surface tensions of components 1 as well as 2 that are associated with their components as per expressions in Equation 7.6 and Equation 7.7:

$$\gamma_{\text{LV}}\left(1+\cos\theta\right) = 2\sqrt{\gamma_{\text{SV}}^d \cdot \gamma_{\text{LV}}^d} + 2\sqrt{\gamma_{\text{SV}}^P \cdot \gamma_{\text{LV}}^P} \tag{7.6}$$

$$\gamma_{\text{SV}} = \gamma_{\text{SV}}^d + \gamma_{\text{SV}}^P \tag{7.7}$$

Where the liquid-vapor, as well as solid-vapor interfaces (surfaces), have been denoted by the subscripts LV or SV, γ^d denotes the surface tension of the nonpolar component, while γ^P denotes the surface tension of the polar component [77].

This system identifies the thermodynamic equilibrium localization for nonreactive, noncompatibilized immiscible binary-component mixtures with nonreactive reinforcing carbon fillers.

Surface tensions (or energies at the surface) of various polymeric matrices can be obtained from the literature and textbooks. Nevertheless, it must be noted that no rankings are prevalently named, and as such, the true surface tension might very well differ from these kinds of estimates as a result of processing or additives at hand in industrial situations. Surface tensions at the operating temperature must be used to compute consistent values in the particular instance of melt blended (nano)composites. These can be measured (for example, utilizing the pendant drop technique) as well as utilizing temperature coefficient, which also is available in tables (see for example [78]) as well as textbooks.

Furthermore, it is hard to ascertain dependable values of surface tension as well as its components in carbon nanomaterials, and values in the literature vary significantly. For instance, the overall CNTs surface tensions are given to range from 27.8 $mN\cdot m^{-1}$ to 114 $mN\cdot m^{-1}$ with polarity values ranging from 0 to 59% [79–82] as a result of diverse features of the explored CNTs as well as distinct analytical techniques. The challenge in forecasting dependable values is aggravated if surface modifications of the CNTs were accomplished to improve polymer compatibility.

The carefully calibrated localization of CNTs clearly relies on the accuracy used for calculations as well as the equations used. Under certain circumstances, thermodynamic predictions agree well with the carbon nanomaterial fillers localization, as reported in the literature [83, 84]. CNT has been predicted and shown to preferential localize in the PET constituent of PET/PVDF/CNT composite materials [84]. Correspondingly, CNTs have been discovered to be predominantly positioned in the poly(propylene carbonate) (PPC) portion of PPC/PLA/MWCNT blend composites based on minimum interfacial energy. This was attributed to PPC/MWCNT having lower interfacial tension than PLA/MWCNT [85]. Furthermore, MWCNTs have always been identified in the PC constituent in melt blended poly(styrene-co-acrylonitrile) (SAN)/PC, even when the surface tension, as well as polarity contrast between the two polymeric materials, are very small [86–88].

The inclusion of nanoparticles in immiscible polymeric blends could influence the overall properties of the blends, especially their phase separation. The addition of nanoscale particles to partially miscible polymer blends results in remarkable behavior, such as changing phase-separation temperatures, changing the form of the phase diagram, or else changing phase separation kinetics. In this regard, the phase boundary has been shown to be pushed up or down depending on the interaction between nanoparticles and polymer mixtures [89]. Lipatov et al. explored chlorinated poly(ethylene)/poly(ethylene-co-vinyl acetate) (EVA) mixtures with silica nanoparticles and found that the temperature of phase separation increased or decreased as per particle concentration [90]. They showed that at both 5% and 10% loading the phase separation rates are significantly reduced by adding filler. *R*(q) values for systems containing with 5 wt.% nanofiller (fumed silica) were lower than those without nanofiller but higher compared to systems loaded with 10% filler, as shown in Figure 7.8. This finding may lend credence to the notion that, in a given situation, they detected phase separation in a blend of higher-molecular-weight components [90]. One of the theories involves unique interactions as well as favored adsorption on the filler by one of the components of the blend. Another possible explanation is that low (or high) molar mass fractions selectively adsorb to the nanofiller surface, altering the local molar mass distribution [90].

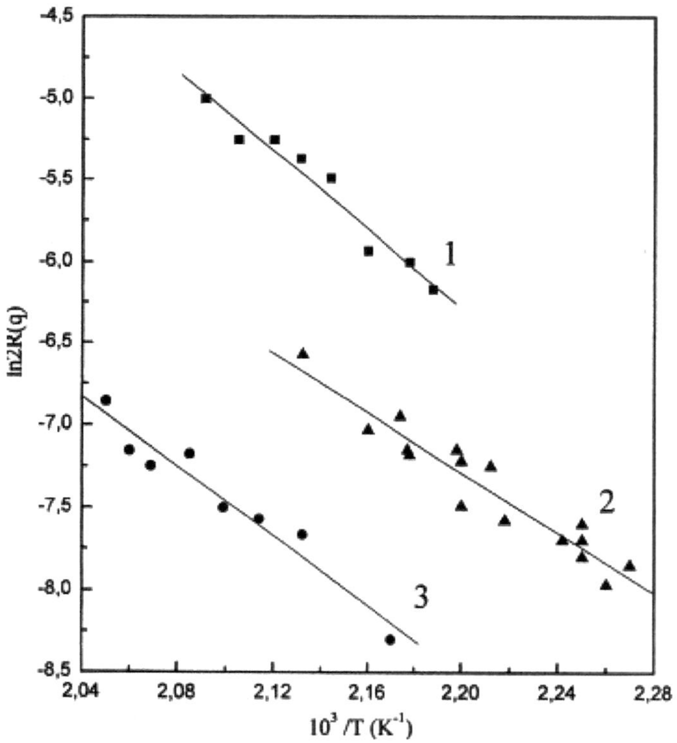

Figure 7.8 Dependence of ln2$R(q)$ on $1/T$ for unfilled (■) and filled 5% (▲) and 10% (●) Aerosil for the blend CPE–EVA at the composition 60/40. Reproduced with permission from [90]. Copyright 2002, Elsevier Science Ltd

In their study, Du et al. [89] investigated the influence of $nSiO_2$ filler on the phase-separation performance of binary polymeric mixtures of PMMA and poly(styrene-co-acrylonitrile) where time-resolved small-angle light scattering along with dynamic rheological measurements were utilized. They showed that the influence of $nSiO_2$ on the phase-separation performance of a PMMA@SAN blend was evidently dependent on the blend matrix composition. When $nSiO_2$ is added, the cloud point (T_c) and binodal temperature (T_b) increase slightly in 60/40 PMMA@SAN and decrease sharply in 30/70 PMMA@SAN. Nevertheless, the addition of $nSiO_2$ affected only a minor increase in spinodal temperatures (T_s) for all PMMA@SAN formulations. As a result, the fillers expand the metastable area of 30/70 PMMA@SAN reinforced with 5 wt.% SiO_2 with a large amplitude, whereas the same region of 60/40 filled PMMA/SAN barely shifts. The selective PMMA absorption on the SiO_2 surface causes a difference in the composition of the surface coating, as well as the polymeric matrices, particularly for 30/70 PMMA/SAN, reinforced with 5% $nSiO_2$ [89]. The phase separation of the surface sheet in such a system occurs via a spinodal decomposition (SD) process at comparatively low annealing temperatures. The $nSiO_2$ particles then perform as nuclei, causing the matrix to phase separate according to the nucleation and growth (NG) process [89].

Mohsenpour et al. used Flory–Huggins theory as well as the Hansen solubility parameters to investigate theoretical binodal boundaries for ternary as well as quaternary blends, which were confirmed using cloud titration experiments. They discovered that adding GO to PES@PVDF blend casting dope solutions enhances inherent thermodynamic unsteadiness, resulting in thinner preferential top layers as well as larger porosities, right up until the stage at which the high viscosity of the GO-comprising solutions prevents solvent-to-nonsolvent demixing. The maximum porosity, as well as pure water flux (PWF), were observed at 0.1 wt.% GO in PES@GO@DMSO and 0.3 wt.% GO for PES@GO/Cyrene and PVDF@GO systems [5]. In their study, they found that the PES@GO membranes cast from DMSO or the PES@GO films cast using DMSO as well as DMF, respectively, showed a decrease in porosity and pure water flux (PWF) when the GO wt.% was raised beyond 0.1 wt.% as well as beyond 0.3 wt.%, respectively. Casting solutions fabricated from lower-molecular-weight PVDF display a wider miscibility gap as well as a reduced viscosity. On the morphology of membranes, these two parameters influence each other. In cross-sectioned SEM images (Figure 7.9) for entire PVDF mixed matrix membranes (MMMs), voids could be observed when 0.3 wt.% GO was incorporated. The topmost layer thickness declined, the porosity, as well as pore size increased; also the porosity, as well as pore size increased. Then again, PVDF/rGO membranes show the opposite pattern, as the thickness of the layer rises with increasing rGO wt.%. In addition, rGO, which is known to have a higher hydrophobicity than GO affected a decline in the affinity of water, as well as the casting dope solution. Ternary phase diagrams remain helpful for forecasting temperature as well as thickness of phase inversion films. They finally suggested that their developed model must be altered to compute the optimal additive concentration for causing upper-limit porosity in the initial polymer solution.

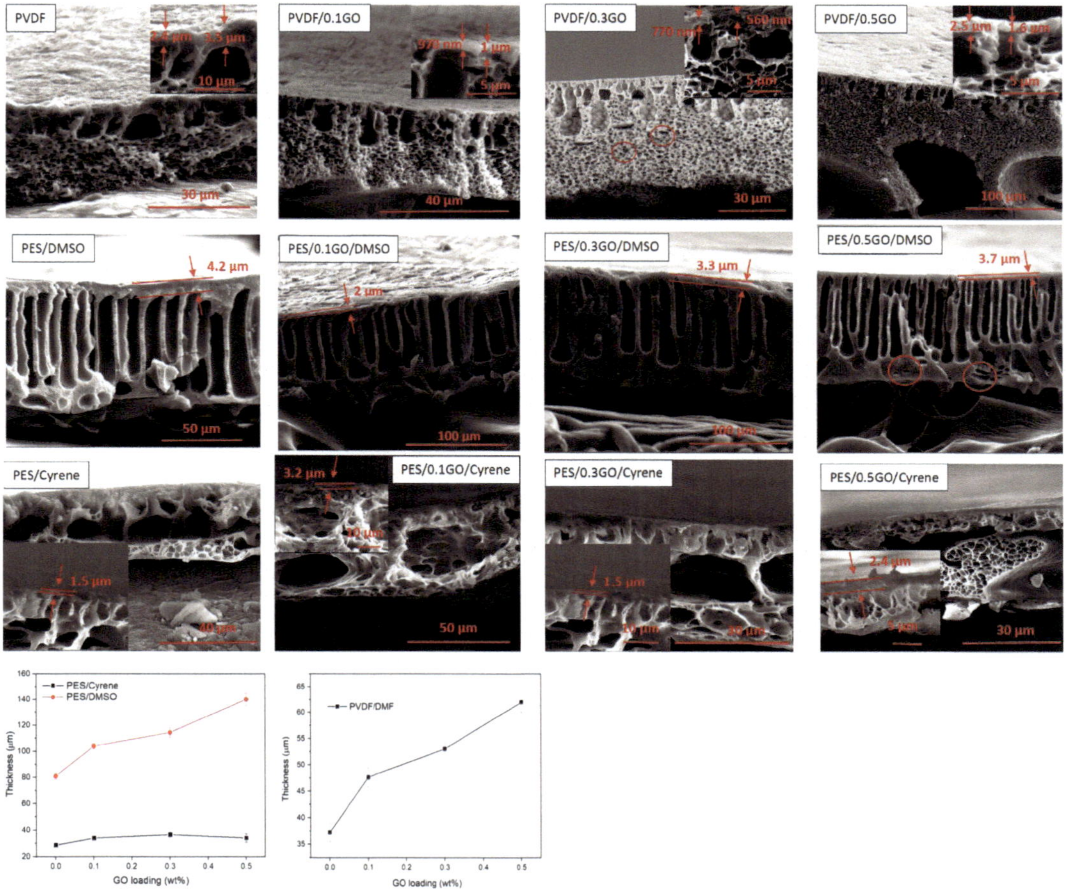

Figure 7.9 Cross-sectional SEM images of the prepared membranes and the thickness of the membrane versus GO loading (wt%). Higher magnification SEM images of PVDF and PES/Cyrene MMMs inset. Reproduced with copyright permission from [5]. Copyright 2011, Elsevier Science Ltd

Again, in a novel study by Zheng et al. [91], COOH@CNTs were effectively integrated into EP@PSF10 as well as EP@PSF20 nanocomposite. Their FTIR spectra revealed interaction amongst COOH@CNT and EP@PSF upon epoxy curing, as well as phase separation of PSF. Because of its higher affinity towards epoxy, COOH@CNTs were preferentially dispersed inside the epoxy phase in both EP@PSF10@CNT-COOH0.5 as well as EP@PSF20@CNT-COOH0.5. Furthermore, the inclusion of COOH@CNTs had no effect on the RIPS mechanisms in either nanocomposite. They did, however, have a substantial influence on the final phase shape, with the PSF dispersed phase in EP@PSF10@CNT-COOH0.5 and the epoxy dispersed phase in EP@PSF20@CNT-COOH0.5 being smaller than those void of CNT-COOHs [91]. Though reaction-induced phase separation should emerge during epoxy curing in the EP@PSF blend, they hypothesized that two-phase separation mechanisms (nucleation growth (NG) as well as spinodal demixing (SD)) might occur. However, with low polysulfone (PSF) content, phase separation followed the NG process, as well as the nanocomposite exhibited a sea-island morphology of PSF microspheres haphazardly scattered in the uninterrupted EP phase. Whereas if PSF wt.% is high enough, a phase-inversion morphology deriving from the SD mechanism might be developed, consisting of epoxy domains as the dispersed phase and PSF as the continuous matrix. Furthermore addition of CNT-COOHs complicated the phase-separation, whereas CNT-COOHs might favor dispersing inside one phase thermodynamically [91]. The inclusion of COOH-CNTs could also influence kinetic factors like viscosity. Figure 7.10 shows SEM as well as HRTEM pictures of EP@PSF10 as well as EP@PSF10@CNT-COOH0.5 fracture morphologies. PSF molecules (red arrows), as well as voids generated by particulate debonding, could be observed here, while the epoxy phase acts as the matrix (blue arrows). The development of said sea-island architecture suggests that RIPS used the NG process. Introducing COOH-CNTs significantly decreases the size of the PSF spheres as well as voids, so there seem to be no gaps in between PSF microspheres as well as an epoxy matrix (Figure 7.10(b,d)). Because of the integration of COOH-CNTs into epoxy, the high viscosity of the trisystem might well have hindered the nucleation as well as the development of PSF in the nanocomposite [27]. As a result, the majority of the PSF was retained within the epoxy web, resulting in smaller PSF spheres. The absence of crevices amongst PSF microspheres with epoxy was attributed to COOH@CNTs, which served a compatibilization role, boosting epoxy compatibility with phase-separated PSF particles (see Figure 7.10(b)). Better interfacial compatibility additionally promoted the composite mechanical performance [91].

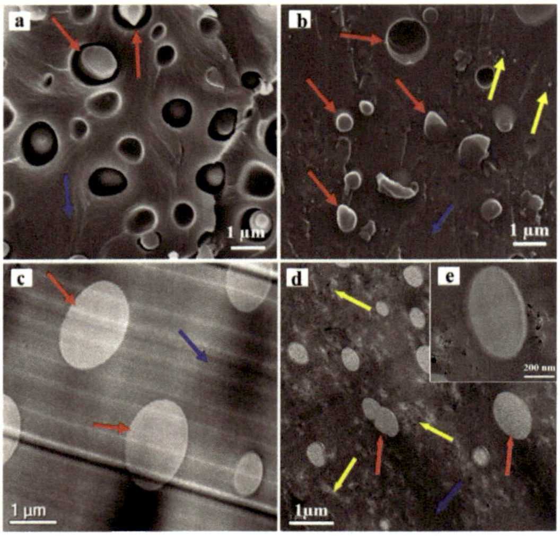

Figure 7.10
SEM and TEM images of (a, c) EP/PSF10 and (b, d, e) EP/PSF10/CNT-COOH0.5. Reproduced with permission from [91]. Copyright 2018, Elsevier Science Ltd

Similar observations, as postulated by Zheng et al., have been reported in recent reports, especially for CNTs [92–96] and graphene-based polymer blended nanocomposites [97, 98].

Depending on the amount of filler used in a polymer blend with LCST, it either raises or lowers the phase-separation temperature. The phase-separation temperature rises as the filler concentration rise to the point where both matrix components go into the state of interfacial layers (which are fully influenced by surface field forces). The shift in the overall thermodynamic interaction constraint in the ternary blends of polymer@polymer@filler explains this phenomenon. At lesser filler concentrations, the polymeric blend components may be reallocated based on their molecular weights between the reinforcing nanofiller surface (within the layer border) and the bulk, which may reduce the phase-separation temperature.

■ 7.3 Conclusions

Manufacturing strategies have incredible significance since they can influence the uniform blending, morphology, and in this way, ultimate properties of the polymeric blends as well as their composites. Techniques such as melt mixing, electrospinning, solvent mixing, and in situ polymerization are comprehensively applied to create polymeric blends as well as their (nano)composites.

When it comes to polymer blends, the main concerns are ease of handling, new preparation technologies, blend structural morphology stability, suitability for innovative advanced applications, as well as recycling. Compatibility opens up new avenues for polymer technology as well as applications because two polymers with excellent intrinsic properties though incompatible, immiscible, or nonreactive, could be used by adapting them into blends with the aid of compatibilizers, resulting in reduced phase-separation. As stated in the preceding sections, thermodynamic miscibility parameters are unavoidable in compatibilization.

Knowing the theoretical foundations and characteristics of the polymeric system as well as the influence of processing approaches is critical owing to the opportunity of key priority characteristics like great elasticity, fantastic toughness, lightweight, transparency, low cost, easy processing, and the capability to satisfy customer requirements. Herein, this chapter attempted to elucidate the fundamental concepts of the influence of the approach to the polymer blend preparation on the properties of the final product.

References

[1] Li, W. Effect of silica nanoparticles on the morphology of polymer blends. *Technische Universiteit Eindhoven*, 1-159 (2011)

[2] Dong, J., Ozaki, Y. FTIR and FT-Raman studies of partially miscible poly(methyl methacrylate)/poly(4-vinylphenol) blends in solid states. *Macromolecules* 30, 286–292 (1997)

[3] Tang, Z. et al. Phase transition behavior of poly(vinylidene fluoride) in a blend with poly(butylene adipate) at high temperature. *Polymer* 194, 122409 (2020)

[4] Jin, X., Li, W., Liu, Y., Gan, W. Self-constructing thermal conductive filler network via reaction-induced phase separation in BNNSs/epoxy/polyetherimide composites. *Composites, Part A* 130, 105727 (2020)

[5] Mohsenpour, S., Leaper, S., Shokri, J., Alberto, M., Gorgojo, P. Effect of graphene oxide in the formation of polymeric asymmetric membranes via phase inversion. *J. Membr. Sci.* 641, 119924 (2022)

[6] Park, S.J., Bates, F.S., Dorfman, K.D. Alternating gyroid in block polymer blends. *ACS Macro Lett.* 11, 643–650 (2022)

[7] Bucher, T., Filiz, V., Abetz, C., Abetz, V. Formation of thin, isoporous block copolymer membranes by an upscalable profile roller coating process – A promising way to save block copolymer. *Membranes* 8, 57 (2018)

[8] Chede, S., Griffiths, P., Escobar, I.C., Harris, T.A. Does casting method matter in filtration membranes? A comparison in performance between doctor blade and slot-die extruded polymeric membranes. *J. Appl. Polym. Sci.* 135, 45563 (2018)

[9] Wang, C. et al. Nonsolvent effects on morphology of cellulose acetate films prepared by dry-cast process. *J. Macromol. Sci., Part B: Phys.* 51, 2266–2275 (2012)

[10] Bhamidipati, K., Didari, S., Harris, T.A. Experimental study on air entrainment in slot die coating of high-viscosity, shear-thinning fluids. *Chem. Eng. Sci.* 80, 195–204 (2012)

[11] Chede, S. et al. Desalination using low biofouling nanocomposite membranes: From batch-scale to continuous-scale membrane fabrication. *Desalination* 451, 81–91 (2019)

[12] Abeykoon, N.C. et al. Novel binder-free electrode materials for supercapacitors utilizing high surface area carbon nanofibers derived from immiscible polymer blends of PBI/6FDA-DAM: DABA. *RSC Adv.* 7, 20947–20959 (2017)

[13] Dong, X. et al. Eco-friendly solvents and their mixture for the fabrication of polysulfone ultrafiltration membranes: An investigation of doctor blade and slot die casting methods. *J. Membr. Sci.* 614, 118510 (2020)

[14] Guillen, G.R., Pan, Y., Li, M., Hoek, E.M. Preparation and characterization of membranes formed by nonsolvent induced phase separation: a review. *Ind. Eng. Chem. Res.* 50, 3798–3817 (2011)

[15] Zhou, Y. et al. Fabrication and electrical characterization of polyaniline-based nanofibers with diameter below 30 nm. *Appl. Phys. Lett.* 83, 3800–3802 (2003)

[16] Angammana, C.J., Jayaram, S.H. Analysis of the effects of solution conductivity on electrospinning process and fiber morphology. *IEEE Trans. Ind. Appl.* 47, 1109–1117 (2011)

[17] Tan, E., Ng, S., Lim, C. Tensile testing of a single ultrafine polymeric fiber. *Biomaterials* 26, 1453–1456 (2005)

[18] Dabirian, F., Hosseini Ravandi, S., Pishevar, A. Investigation of parameters affecting PAN nanofiber production using electrical and centrifugal forces as a novel method. *Curr. Nanosci.* 6, 545–552 (2010)

[19] Zhang, Y., Huang, Z.-M., Xu, X., Lim, C.T., Ramakrishna, S. Preparation of core-shell structured PCL-r-gelatin bi-component nanofibers by coaxial electrospinning. *Chem. Mater.* 16, 3406–3409 (2004)

[20] Dalton, P. D., Klee, D., Möller, M. Electrospinning with dual collection rings. *Polymer* 46, 611–614 (2005)

[21] Nayani, K. et al. Electrospinning combined with nonsolvent-induced phase separation to fabricate highly porous and hollow submicrometer polymer fibers. *Ind. Eng. Chem. Res.* 51, 1761–1766 (2012)

[22] Khajavi, R., Abbasipour, M. Electrospinning as a versatile method for fabricating coreshell, hollow and porous nanofibers. *Sci. Iran.* 19, 2029–2034 (2012)

[23] Dayal, P., Liu, J., Kumar, S., Kyu, T. Experimental and theoretical investigations of porous structure formation in electrospun fibers. *Macromolecules* 40, 7689–7694 (2007)

[24] Matsuyama, H., Takida, Y., Maki, T., Teramoto, M. Preparation of porous membrane by combined use of thermally induced phase separation and immersion precipitation. *Polymer* 43, 5243–5248 (2002)

[25] Kuo, C.-Y., Lin, H.-N., Tsai, H.-A., Wang, D.-M., Lai, J.-Y. Fabrication of a high hydrophobic PVDF membrane via nonsolvent induced phase separation. *Desalination* 233, 40–47 (2008)

[26] Lee, H. J., Jung, B., Kang, Y. S., Lee, H. Phase separation of polymer casting solution by nonsolvent vapor. *J. Membr. Sci.* 245, 103–112 (2004)

[27] McCann, J. T., Marquez, M., Xia, Y. Highly porous fibers by electrospinning into a cryogenic liquid. *J. Am. Chem. Soc.* 128, 1436–1437 (2006)

[28] Wei, M., Kang, B., Sung, C., Mead, J. Core-sheath structure in electrospun nanofibers from polymer blends. *Macromol. Mater. Eng.* 291, 1307–1314 (2006)

[29] Yarin, A. Coaxial electrospinning and emulsion electrospinning of core-shell fibers. *Polym. Adv. Technol.* 22, 310–317 (2011)

[30] Wang, C., Hsiue, T.-T. Core-shell fibers electrospun from phase-separated blend solutions: fiber formation mechanism and unique energy dissipation for synergistic fiber toughness. *Biomacromolecules* 18, 2906–2917 (2017)

[31] Chiu, Y. J., Tseng, H. F., Lo, Y. C., Wu, B. H., Chen, J. T. Plateau-Rayleigh instability morphology evolution (PRIME): From electrospun core-shell polymer fibers to polymer microbowls. *Macromol. Rapid Commun.* 38, 1600689 (2017)

[32] Zhang, J.-F. et al. Electrospun core-shell structure nanofibers from homogeneous solution of poly(ethylene oxide)/chitosan. *Macromolecules* 42, 5278–5284 (2009)

[33] Chen, M. et al. Thermo-responsive core-sheath electrospun nanofibers from poly(*N*-isopropylacrylamide)/polycaprolactone blends. *Chem. Mater.* 22, 4214–4221 (2010)

[34] Jiang, Y. et al. Fabrication of core-shell nanofibers by single capillary electrospinning combined with vapor induced phase separation. *New J. Chem.* 37, 2917–2924 (2013)

[35] Xu, X. et al. Preparation of core-sheath composite nanofibers by emulsion electrospinning. *Macromol. Rapid Commun.* 27, 1637–1642 (2006)

[36] Bazilevsky, A. V., Yarin, A. L., Megaridis, C. M. Co-electrospinning of core-shell fibers using a single-nozzle technique. *Langmuir* 23, 2311–2314 (2007)

[37] Jo, E. et al. Core-sheath nanofibers containing colloidal arrays in the core for programmable multi-agent delivery. *Adv. Mater.* 21, 968–972 (2009)

[38] Yang, Y. et al. Structural stability and release profiles of proteins from core-shell poly(DL-lactide) ultrafine fibers prepared by emulsion electrospinning. *J. Biomed. Mater. Res., Part A* 86, 374–385 (2008)

[39] Heriot, S. Y., Jones, R. A. An interfacial instability in a transient wetting layer leads to lateral phase separation in thin spin-cast polymer-blend films. *Nat. Mater.* 4, 782–786 (2005)

[40] Liu, T., Ozisik, R., Siegel, R. W. Phase separation and surface morphology of spin-coated films of polyetherimide/polycaprolactone immiscible polymer blends. *Thin Solid Films* 515, 2965–2973 (2007)

[41] Chang, L.L., Woo, E.M. Surface morphology and Flory-Huggins interaction strength in UCST blend system comprising poly(4-methyl styrene) and isotactic polystyrene. *Polymer* 44, 1711–1719 (2003)

[42] Bergues, B. et al. Phase decomposition in polymer blend films cast on homogeneous substrates modified by self-assembled monolayers. *Vacuum* 63, 297–305 (2001)

[43] Cyganik, P. et al. Phase decomposition in polymer blend films cast on substrates patterned with self-assembled monolayers. *Vacuum* 63, 307–313 (2001)

[44] Cyganik, P., Budkowski, A., Raczkowska, J., Postawa, Z. AFM/LFM surface studies of a ternary polymer blend cast on substrates covered by a self-assembled monolayer. *Surf. Sci.* 507–510, 700–706 (2002)

[45] Ebbens, S. et al. In situ imaging and height reconstruction of phase separation processes in polymer blends during spin coating. *ACS Nano* 5, 5124–5131 (2011)

[46] Buxton, G., Clarke, N. Ordering polymer blend morphologies via solvent evaporation. *Europhys. Lett.* 78, 56006 (2007)

[47] Guo, X. et al. A new strategy of lithography based on phase separation of polymer blends. *Sci. Rep.* 5, 1–12 (2015)

[48] Mokarian-Tabari, P. et al. Quantitative evaluation of evaporation rate during spin-coating of polymer blend films: control of film structure through defined-atmosphere solvent-casting. *Eur. Phys. J. E: Soft Matter Biol. Phys.* 33, 283–289 (2010)

[49] Karim, A. et al. Phase separation of ultrathin polymer-blend films on patterned substrates. *Phys. Rev. E* 57, R6273–R6276 (1998)

[50] Zhang, W., Deodhar, S., Yao, D. Geometrical confining effects in compression molding of co-continuous polymer blends. *Ann. Biomed. Eng.* 38, 1954–1964 (2010)

[51] Fortelný, I., Lapčíková, M., Lednický, F., Starý, Z., Kruliš, Z. Nonuniformity of phase structure in immiscible polymer blends. *Polym. Eng. Sci.* 48, 564–571 (2008)

[52] Frone, A.N. et al. Morpho-structural, thermal and mechanical properties of PLA/PHB/Cellulose biodegradable nanocomposites obtained by compression molding, extrusion, and 3D printing. *Nanomaterials* 10, 51 (2019)

[53] Zhang, W., Li, M., Deodhar, S., Yao, D., Zhou, J.G. In *ASME 2010 International Manufacturing Science and Engineering Conference*, October 12–15, 2010 Erie, Pennsylvania, USA, Vol. 1, 7–11 (2010)

[54] Bose, S. et al. Phase separation as a tool to control dispersion of multiwall carbon nanotubes in polymeric blends. *ACS Appl. Mater. Interfaces* 2, 800–807 (2010)

[55] Clegg, P. Fluid-bicontinuous gels stabilized by interfacial colloids: low and high molecular weight fluids. *J. Phys.: Condens. Matter* 20, 113101 (2008)

[56] Murakami, R., Bismarck, A. Particle-stabilized materials: dry oils and (polymerized) non-aqueous foams. *Adv. Funct. Mater.* 20, 732–737 (2010)

[57] Sundararaj, U., Macosko, C. Drop breakup and coalescence in polymer blends: the effects of concentration and compatibilization. *Macromolecules* 28, 2647–2657 (1995)

[58] Gubbels, F. et al. Design of electrical composites: determining the role of the morphology on the electrical properties of carbon black filled polymer blends. *Macromolecules* 28, 1559–1566 (1995)

[59] Vermant, J., Cioccolo, G., Golapan Nair, K., Moldenaers, P. Coalescence suppression in model immiscible polymer blends by nano-sized colloidal particles. *Rheol. Acta* 43, 529–538 (2004)

[60] Zhang, Q., Yang, H., Fu, Q. Kinetics-controlled compatibilization of immiscible polypropylene/polystyrene blends using nano-SiO_2 particles. *Polymer* 45, 1913–1922 (2004)

[61] Li, L. et al. Kinetically trapped co-continuous polymer morphologies through intraphase gelation of nanoparticles. *Nano Letts.* 11, 1997–2003 (2011)

[62] Khatua, B., Lee, D.J., Kim, H.Y., Kim, J.K. Effect of organoclay platelets on morphology of nylon-6 and poly(ethylene-ran-propylene) rubber blends. *Macromolecules* 37, 2454–2459 (2004)

[63] Si, M. et al. Compatibilizing bulk polymer blends by using organoclays. *Macromolecules* 39, 4793–4801 (2006)

[64] Vandebril, S., Vermant, J., Moldenaers, P. Efficiently suppressing coalescence in polymer blends using nanoparticles: role of interfacial rheology. *Soft Matter* 6, 3353–3362 (2010)

[65] Fenouillot, F., Cassagnau, P., Majesté, J.-C. Uneven distribution of nanoparticles in immiscible fluids: Morphology development in polymer blends. *Polymer* 50, 1333–1350 (2009)

[66] Kelnar, I., Khunová, V., Kotek, J., Kaprálková, L. Effect of clay treatment on structure and mechanical behavior of elastomer-containing polyamide 6 nanocomposite. *Polymer* 48, 5332–5339 (2007)

[67] Huitric, J., Ville, J., Médéric, P., Moan, M., Aubry, T. Rheological, morphological and structural properties of PE/PA/nanoclay ternary blends: Effect of clay weight fraction. *J. Rheol.* 53, 1101–1119 (2009)

[68] Hong, J.S., Namkung, H., Ahn, K.H., Lee, S.J., Kim, C. The role of organically modified layered silicate in the breakup and coalescence of droplets in PBT/PE blends. *Polymer* 47, 3967–3975 (2006)

[69] Elias, L., Fenouillot, F., Majesté, J.-C., Cassagnau, P. Morphology and rheology of immiscible polymer blends filled with silica nanoparticles. *Polymer* 48, 6029–6040 (2007)

[70] Gong, Y., Leal, L.G. Role of symmetric grafting copolymer on suppression of drop coalescence. *J. Rheol.* 56, 397–433 (2012)

[71] Martin, J.D., Velankar, S.S. Effects of compatibilizer on immiscible polymer blends near phase inversion. *J. Rheol.* 51, 669–692 (2007)

[72] Thareja, P., Moritz, K., Velankar, S.S. Interfacially active particles in droplet/matrix blends of model immiscible homopolymers: Particles can increase or decrease drop size. *Rheol Acta* 49, 285–298 (2010)

[73] Filippone, G., Salzano de Luna, M. A unifying approach for the linear viscoelasticity of polymer nanocomposites. *Macromolecules* 45, 8853–8860 (2012)

[74] Pawar, A.B., Caggioni, M., Hartel, R.W., Spicer, P.T. Arrested coalescence of viscoelastic droplets with internal microstructure. *Faraday Discuss.* 158, 341–350 (2012)

[75] Nagarkar, S.P., Velankar, S.S. Morphology and rheology of ternary fluid-fluid-solid systems. *Soft Matter* 8, 8464–8477 (2012)

[76] Sumita, M., Sakata, K., Asai, S., Miyasaka, K., Nakagawa, H. Dispersion of fillers and the electrical conductivity of polymer blends filled with carbon black. *Polym. Bull.* 25, 265–271 (1991)

[77] Wu, S. *Polymer Interface and Adhesion*. Marcel Dekker, New York (1982)

[78] http://www.surface-tension.de/solid-surface-energy.htm [accessed Dec 12, 2022]

[79] Barber, A.H., Cohen, S.R., Wagner, H.D. Static and dynamic wetting measurements of single carbon nanotubes. *Phys. Rev. Lett.* 92, 186103 (2004)

[80] Menzel, R., Lee, A., Bismarck, A., Shaffer, M.S. Inverse gas chromatography of as-received and modified carbon nanotubes. *Langmuir* 25, 8340–8348 (2009)

[81] Stöckelhuber, K.W., Das, A., Jurk, R., Heinrich, G. Contribution of physico-chemical properties of interfaces on dispersibility, adhesion and flocculation of filler particles in rubber. *Polymer* 51, 1954–1963 (2010)

[82] Nuriel, S., Liu, L., Barber, A., Wagner, H. Direct measurement of multiwall nanotube surface tension. *Chem. Phys. Lett.* 404, 263–266 (2005)

[83] Ko, S.W., Gupta, R.K., Bhattacharya, S.N., Choi, H.J. Rheology and physical characteristics of synthetic biodegradable aliphatic polymer blends dispersed with MWNTs. *Macromol. Mater. Eng.* 295, 320–328 (2010)

[84] Wu, M., Shaw, L.L. On the improved properties of injection-molded, carbon nanotube-filled PET/PVDF blends. *J. Power Sources* 136, 37–44 (2004)

[85] Park, D.H., Kan, T.G., Lee, Y.K., Kim, W.N. Effect of multi-walled carbon nanotube dispersion on the electrical and rheological properties of poly(propylene carbonate)/poly(lactic acid)/multi-walled carbon nanotube composites. *J. Mater. Sci.* 48, 481–488 (2013)

[86] Gültner, M., Göldel, A., Pötschke, P. Tuning the localization of functionalized MWCNTs in SAN/PC blends by a reactive component. *Compos. Sci. Technol.* 72, 41–48 (2011)

[87] Göldel, A., Kasaliwal, G.R., Pötschke, P., Heinrich, G. The kinetics of CNT transfer between immiscible blend phases during melt mixing. *Polymer* 53, 411–421 (2012)

[88] Göldel, A., Kasaliwal, G., Pötschke, P. Selective localization and migration of multiwalled carbon nanotubes in blends of polycarbonate and poly(styrene-acrylonitrile). *Macromol. Rapid Commun.* 30, 423–429 (2009)

[89] Du, M., Wu, Q., Zuo, M., Zheng, Q. Filler effects on the phase separation behavior of poly(methyl methacrylate)/poly(styrene-co-acrylonitrile) binary polymer blends. *Eur. Polym. J.* 49, 2721–2729 (2013)

[90] Lipatov, Y.S., Nesterov, A., Ignatova, T., Nesterov, D. Effect of polymer–filler surface interactions on the phase separation in polymer blends. *Polymer* 43, 875–880 (2002)

[91] Zheng, N. et al. Effects of carboxylated carbon nanotubes on the phase separation behaviour and fracture-mechanical properties of an epoxy/polysulfone blend. *Compos. Sci. Technol.* 159, 180–188 (2018)

[92] Wang, P. et al. Effect of hydroxyl and carboxyl-functionalized carbon nanotubes on phase morphology, mechanical and dielectric properties of poly(lactide)/poly(butylene adipate-co-terephthalate) composites. *Int. J. Bio. Macromol.* 206, 661–669 (2022)

[93] Ganji, P., Nazari, S., Zinatizadeh, A.A., Zinadini, S. Chitosan-wrapped multi-walled carbon nanotubes (CS/MWCNT) as nanofillers incorporated into nanofiltration (NF) membranes aiming at remarkable water purification. *J. Water Proc. Eng.* 48, 102922 (2022)

[94] Gholami, S. et al. Impact of a new functionalization of multiwalled carbon nanotubes on antifouling and permeability of PVDF nanocomposite membranes for dye wastewater treatment. *Chemosphere* 294, 133699 (2022)

[95] Sims, C.M., Fagan, J.A. Surfactant chemistry and polymer choice affect single-wall carbon nanotube extraction conditions in aqueous two-polymer phase extraction. *Carbon* 191, 215–226 (2022)

[96] Norouzi, A., Kojabad, M.E., Chapalaghi, M., Hosseinkhani, A., Lay, E.N. Polyester-based polyurethane mixed-matrix membranes incorporating carbon nanotube-titanium oxide coupled nanohybrid for carbon dioxide capture enhancement: molecular simulation and experimental study. *J. Mol. Liq.* 119540 (2022)

[97] Hostettler, N., Hubert, P. A phase separation strategy for enhanced toughness self-assembly graphene-network composites. *Compos. Sci. Technol.* 109503 (2022)

[98] Sahan, M.S., Abdullah, E.T. Morphological and electrical properties of polyvinylpyrrolidone/multi-walled carbon nanotubes nanocomposite with graphene. *Basrah J. Sci.* 40(1), 128–137 (2022)

8 Interfacial Modification of Immiscible Polymer Blends

8.1 Introduction

Multiphase polymer structures are of paramount importance because they can be used in different applications [1–3]. When two polymers are mixed via different preparative methods, they can either be single-phase miscible polymer blends or phase-separated polymer blend systems [1]. Miscibility in polymer blend systems is mainly governed by the Gibbs free energy of mixing, as given in Equation 8.1

$$\Delta G_{mix} = \Delta H_{mix} - T\Delta S_{mix} \tag{8.1}$$

The miscibility of two polymers is mainly due to the enthalpy of mixing ΔH_{mix} as compared to the entropy ΔS_{mix} because entropy runs low in large macromolecules. A positive value of the Gibbs free energy ΔG_{mix} usually leads to an immiscible polymer blend system, whilst a negative value results in miscibility. Most polymer blend systems are usually immiscible. Miscibility usually results due to specific interactions which may arise from interfacial bonding, such as dipole-dipole, hydrogen bonding, copolymer repulsion effect, and ion-ion interactions, just to mention a few [4–6]. When polymer blends demix, they can either demix to produce a co-continuous or droplet matrix morphology. The co-continuous morphology is one of the most favored types of morphology because the minor phase can be extracted, and the porous system can be used to develop membranes that can be used for water remediation, scaffolds for biomedical engineering, and separators for lithium batteries. The co-continuous morphology is obtained through the blending of two immiscible polymers or via the demixing of a partially miscible blend system. This morphology is thermodynamically unstable, and with time they can coarsen into a droplet matrix morphology. The microphase separation in demixed polymer blends usually results in the overall system having poor mechanical properties, which is not a good attribute, especially in the fabrication of polymer products for different applications. The thick interface in immiscible polymer blends usually acts as crack initiators when a load is applied, leading to failure. To convert the immiscible polymer blends into useful polymeric products, it is of paramount importance to control and manipulate the phase behavior of the systems. Different methods have been investigated to control the interfacial area of two immiscible polymers, and the use of compatibilizers to lower the thickness of the interface and improve the mechanical properties is one of them. This chapter will mainly focus on different state-of-the-art methods used in manipulating the interface of immiscible polymer blend systems.

■ 8.2 Block Copolymers

The use of block copolymers (BCP) to tune the interfacial area of two immiscible polymers is an effective strategy used in compatibilizing phase-separated polymer blend systems. The presence of the functional groups on the block copolymers leads to a reaction with one component of the polymer blend inducing miscibility. The BCP is designed to be trapped at the interface of the immiscible polymer blend to enhance interfacial interaction. Ding et al. [7] prepared PLA-polyethylene glycol-PLA, a triblock polymer, to compatibilize a polymer blend system of PLA (polylactic acid)/PBAT (polybutylene adipate-co-terephthalate). The polymer system was prepared via the melt mixing method together with the compatibilizer. The triblock copolymers were shown to effectively compatibilize the immiscible polymer blend system as shown by different characterization methods. The use of block copolymers as compatibilizers has faced some challenges due to the controlled polymerization; henceforth, quasi-block copolymers have been utilized. Zheng et al. [8] made use of quasi-block copolymers of 1-vinyl-3-butylimidazole chloride, grafted polyamide 6 (PA6-g-IL(Cl)) to compatibilize the poly(vinylidene fluoride)/polyamide 6 (PVDF/PA6) polymer blend system. Incorporation of the PA6-g-IL(Cl) into the immiscible polymer blend resulted in a change of morphology from sea-island to bicontinuous morphology. This indicated effective compatibilization of the polymer blend system. The effectiveness of compatibilizing the system was investigated by using different characterization methods and by using rheological experiments. It was observed that the addition of the compatibilizer increased both the storage modulus (G') and the complex viscosity (η^*). The compatibilization mechanism was mainly attributed to the interaction between the PVDF macromolecules and the blocks of ionic-liquid-grafted polyamide 6, the polyamide 6 which was ungrafted also entangled with the polyamide 6 chains. Another disadvantage of using block copolymers is that they can result in micelles when the concentration is more than the critical value leading to a new phase which will result in a change in the properties of the polymer blend. Researchers have gone a step further in developing comb-like copolymers, Wang et al. [9] made use of comb-like copolymers to investigate their effectiveness in compatibilizing LLDPE (linear low-density polyethylene)/ PS (polystyrene) immiscible polymer blend system. Graft copolymers can be applied in stabilizing immiscible systems. To improve the properties of poly(3-hydroxybutyrate-co-3-hydroxyhexanoate) (PHBH), Tamiya et al. [10] blended PHBH with PMMA and compatibilized the system with a graft copolymer (PMMA-g-PHBH). The compatibilizer was added to the polymer system to enhance the miscibility and the mechanical properties of the system. It was observed in the study that the domain sizes decreased when PMMA-g-PHBH was added, mainly due to the increased interfacial adhesion between PMMA and PHBH. This improvement in miscibility was due to the graft copolymer, which was able to interact with the two polymers.

■ 8.3 Janus Nanoparticles

It is of paramount importance to control and manipulate the morphology of demixed polymer blends, especially co-continuous morphology. This type of morphology is usually not thermodynamically stable and can coarsen to result in droplet matrix morphology over time. One way to stabilize this type of morphology is by making use of Janus nanoparticles. Because Janus nanoparticles can be localized at the interface, they provide effective stability compared to nanoparticles which will usually be located in a single phase. Janus particles are materials that usually possess two surfaces having different characteristics. These unique properties enable Janus nanoparticles to be used in different applications [11], and one of them is the compatibilization of immiscible blend systems [12]. The asymmetrical structure and amphiphilicity of the Janus nanoparticles enable them to be localized on the interface of immiscible blend systems. Their incorporation in immiscible blends improves the mechanical and other different properties of immiscible blend systems. The advent of Janus nanoparticles has significantly impacted the stabilization of immiscible polymer blend systems. Many researchers have shown a keen interest in developing these unique nanoparticles due to their main advantage such as their dual nature which enables them to have fascinating properties. Janus nanoparticles can be prepared by different methods, and some of the methods have been outlined in Figure 8.1, a detailed outline of synthetic methods can be obtained from a review by Lattuada and Hatton [13].

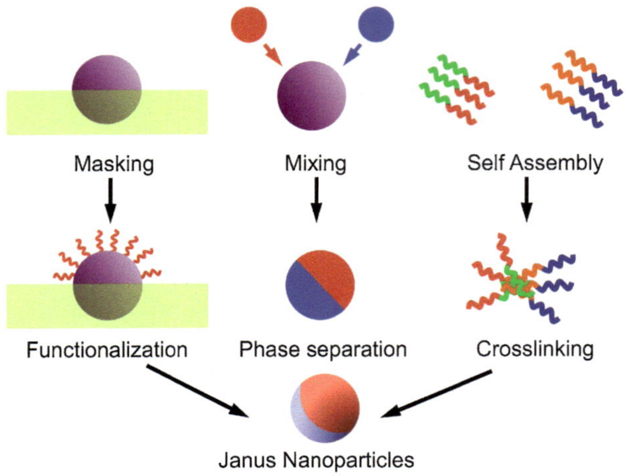

Figure 8.1 Schematic diagram illustrating the three main strategies for the preparation of Janus nanoparticles: masking, phase separation, and self-assembly. Reprinted with permission from [13]. Copyright 2011, Elsevier Science Ltd

Janus nanoparticles (JP) can be localized on the interface of the two polymers, and they can act as effective compatibilizers by lowering the interfacial tension of immiscible polymer blends. In their work, Xu et al. [14] fabricated MPS-SiO_2@PDVB-DM Janus particles, where MPS is 3-(trimethoxysilyl)propyl methacrylate, SiO_2 is silicon dioxide, PDVB is polydivinylbenzene, and DM is dodecyl mercaptan. The JPs were then added to a blended system of liquid isoprene rubber (LIR)/epoxy resin (ER) to compatibilize the polymer blend system [14]. The MPS-SiO_2 moves towards the ER, and then the PDVB-DM is attracted towards the LIR phase leading to the JP being trapped on the interface, as shown in Figure 8.2.

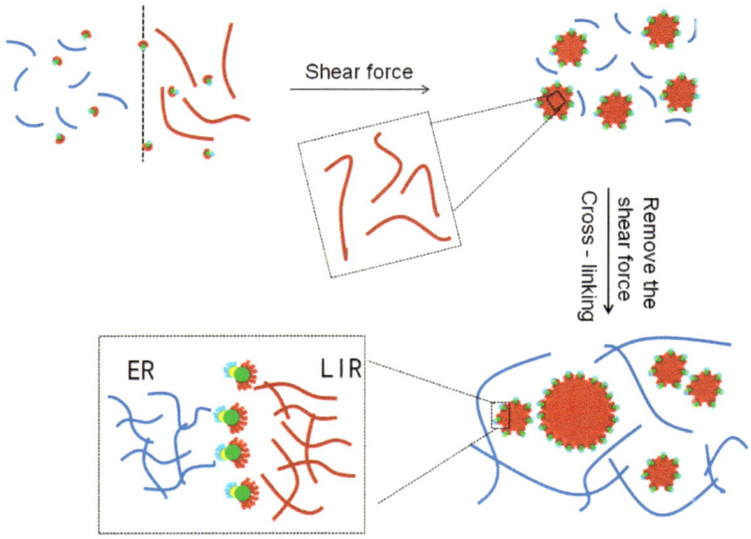

Figure 8.2 Diagram of the orientation of MPS-SiO_2@PDVBDM Janus particles in LIR/ER blends. Reprinted with permission from [14]. Copyright 2017, American Chemical Society

You et al. [15] prepared snowman-like JPs and incorporated them in an immiscible polymer blend system of poly(lactic acid)/polycarbonate [15]. The JPs were fabricated with different shape asymmetries, as illustrated in Figure 8.3.

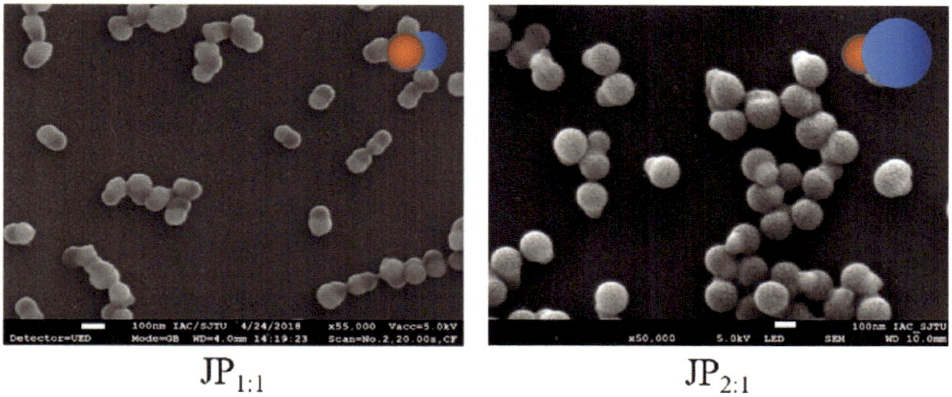

Figure 8.3 Transmission electron microscope images of the snowman-like JP with different shape asymmetries. Reprinted with permission from [15]. Copyright 2018, American Chemical Society

The localization of JPs at the interface of the demixed blend system was attributed to the controlling of the chemical asymmetry. JPs were shown to lower the size of the domains resulting in compatibilization of the system. To obtain maximum compatibilizing effect from different types of compatibilizers, it is of great importance for them to be localized at the interface. The main disadvantage of using block copolymers is that they can move away from the interface when exposed to conditions with high shear stresses. In an endeavor to form compatibilized PS/PMMA, Walther et al. [16] fabricated JPs with both PS and PMMA portions, as shown in Figure 8.4, to compatibilize the immiscible polymer blend system.

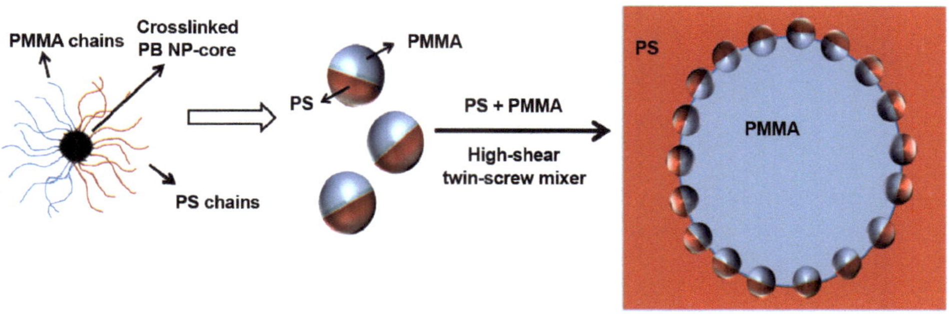

Figure 8.4 Schematic representation of Janus particles and their adsorption at the blend interface of a PS/PMMA blend. Reproduced with permission from [16]. Copyright 2008 American Chemical Society

Walther et al. [16] found that the reduction in the domain size (see Figure 8.5) as increasing amounts of JPs were added further proved that they are good interfacial compatibilizers of an immiscible blend system. Their results showed that the JPs can act as effective compatibilizers even when subjected to high shear conditions. Rheology can be used as an effective tool in determining the effectiveness of Janus nanomaterials in compatibilizing immiscible polymer blend systems. Parpaite et al. [17] made use of Janus hybrid nanoparticles as compatibilizers in a polystyrene/polyamide 6 polymer blend system. The hybrid nanoparticles were found to be localized at the interface and were shown by melt rheology to be effective compatibilizers of PS/PA6 polymer blend system. The reduction in the effective interfacial tension was determined by viscoelastic measurements.

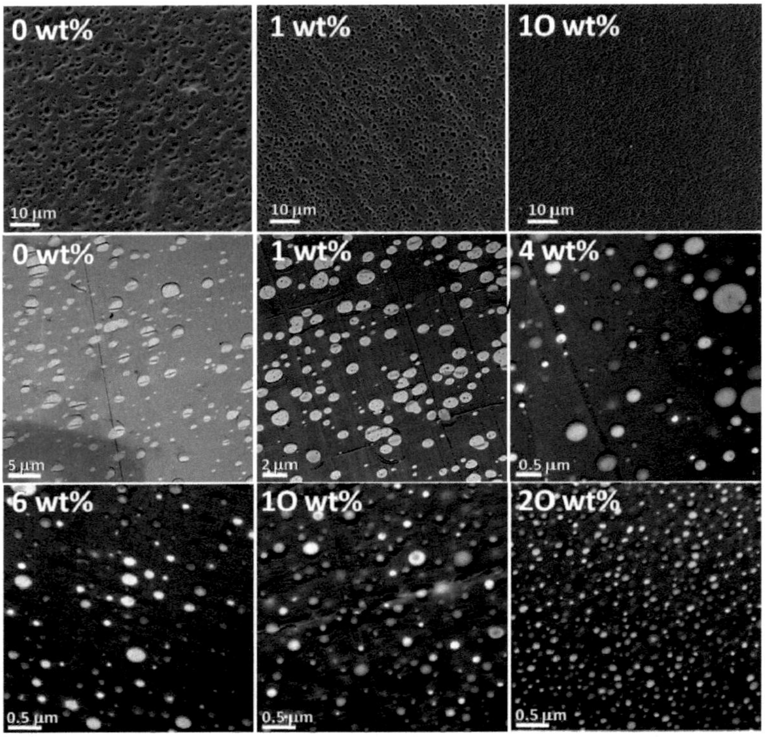

Figure 8.5 Scanning electron microscope images obtained for blends at a PS/PMMA ratio of 6/4 (top row). Nonstained TEM images obtained for blends at a PS/PMMA ratio of 8/2 (center and lower row). The amount of compatibilizer is indicated in each image. Reprinted with permission from [16]. Copyright 2008, American Chemical Society

■ 8.4 Compatibilization by Nanoparticles

Compatibilizing materials are of paramount significance in fabricating multiphasic systems, which can be used in different applications. Different nanomaterials, such as graphene oxide and carbon nanotubes, can be incorporated into the systems to enhance the interfacial adhesion of immiscible polymer blend systems.

8.4.1 Graphene Oxide

Even though copolymers can be used in stabilizing immiscible polymer blend systems, they have several disadvantages that they offer compared to nanomaterials [18, 19]. Copolymers do not improve to a greater extent the stiffness and other mechanical properties of the immiscible blend systems. Copolymers are very difficult to synthesize and as a result they are expensive to produce. Nanomaterials, whether functionalized or nonfunctionalized, can be incorporated into immiscible polymer blend systems to reduce interfacial tension. Graphene oxide (GO) is a nanomaterial that can be used to compatibilize immiscible blend systems. GO is an amphiphile because of the presence of the hydrophilic groups on its edges. This nature means that GO can be used as a surfactant in different applications and can be used as an effective compatibilizer due to its amphiphilic nature. GO was found to improve the compatibility of PA (polyamide)/PPO (polyphenylene oxide) [20]. This was because the GO nanomaterial interacted with both the polymers as it forms hydrogen bonding with PA and also exhibits π-π interactions with PPO. Tong et al. [21] also reported the improved compatibility between polystyrene (PS) and poly(vinylidene fluoride) (PVDF). The GO sheets were located at the interface of the two polymers via water-assisted extrusion. The compatibilization effect was a result of the interaction of GO functional groups and the CF_2 in PVDF and the interaction of the GO sheets with phenyl rings of PS. When subjected to high processing conditions, the amphiphile properties of GO can be lost, resulting in it losing its compatibilizing properties in immiscible polymer blend systems. Ye et al. [22] investigated the compatibilizing effect of GO in the presence of PMMA/PS. The compatibilizing effect of the nanomaterials was shown to decrease when subjected to high processing temperatures due to the removal of polar entities on the GO surface. At low temperatures, the GO can act as a compatibilizer in PMMA/PS because there will be π-π stacking between GO and PS and GO forms hydrogen bonds with PMMA. At high temperatures, the GO then loses its effect of compatibilizing the immiscible polymer blend system. Improvement in mechanical properties of immiscible blends in the presence of GO can be a sign that the nanomaterials are good compatibilizers. When the GO sheets are evenly distributed inside the polymer matrix, an improvement in the mechanical properties can be observed irrespective of the fact that it is localized on the interface or in another phase. GO sheets were observed to enhance the miscibility and the mechanical properties of NR/HDPE even though the nanomaterials were localized in the NR phase [23]. Functionalized GO sheets can also be used in compatibilizing immiscible polymer blend systems. Functionalization of GO leads to better dispersion inside the host polymer matrix. Interfacially localized octadecylamine-functionalized GO was found to improve the miscibility of the polydimethylsiloxane (PDMS)/polyisoprene (PI) immiscible polymer blend system [24]. Polyolefins tend not to interact well with GO nanosheets because they are not polar, and GO is polar. GO can be grafted with nonpolymer polymers to improve its efficiency in compatibilizing nonpolar polymers. Polypropylene-grafted GO (PP-g-GO) was synthesized and was shown to enhance the compatibility of the PP/PPO immiscible polymer blend system [25]. The PP-g-GO migrated to the interface of the two polymers because of the interaction of grafted PP with the PP matrix via intermolecular interactions and due to the GO and PPO interactions.

8.4.2 Carbon Nanotubes

CNTs are nanoparticles that can be used as compatibilizers in immiscible polymer blend systems [26]. The specific localization of CNTs in one of the polymer phases is of great significance because their localization usually affects the overall properties of the polymer blend system [27]. The localization of CNTs is usually governed by thermodynamics, the mixing protocol, and viscosity. CNTs can induce compatibility in immiscible systems through interaction with both polymers when they are kinetically arrested at the interface, and they can prevent droplet coalescence resulting in increased miscibility. Improved interfacial adhesion can be obtained when the CNTs are chemically functionalized so that they can interact with both polymers, especially when they are at the interface. The incompatibility of polymers usually reduces their chances of being used in different applications. Poly(ε-caprolactone) (PCL) and polylactide (PLA) are two polymers that can be blended and used in many biomedical applications. However, their main limitation is their poor adhesion when mixed. A simple way of improving their compatibility is through the addition of CNTs. Wu et al. [28] used functionalized CNTs to improve the adhesion of this interesting polymer blend system. The nanoparticles were found to be localized both at the interface and in the PCL phase. The interfacial adhesion was improved because the nanoparticles prevented coalescence. Zhao et al. [29] made use of this strategy by chemically functionalizing CNTs with PMMA chains and epoxy groups in a polyvinylidene fluoride (PVDF)/poly(L-lactide) (PLLA) immiscible polymer blend system. The PMMA polymer chains interacted with the PVDF since the two are miscible [6, 30, 31], and the reactive epoxy groups interacted with PLLA. These interactions enabled the functionalized CNTs to be located at the interface and enhanced the interfacial adhesion. The localization of CNTs is a challenge due to poor interfacial stability. This may result in the ineffectiveness of the nanomaterials in improving the adhesion of immiscible polymer blend systems. Different strategies can be implemented to kinetically arrest the nanoparticles at the interface, and one of them is by making use of stereo-complex crystallites [32]. The formation of the crystallites during melt processing may hinder the migration of the CNTs into the thermodynamically favored phase resulting in the nanoparticles being located at the interface.

■ 8.5 Conclusions

It can be summarized that block copolymers, Janus nanoparticles, and different types of nanomaterials such as GO and CNTs can be incorporated into different immiscible polymer blend systems as effective compatibilizers. Block copolymers can be tuned such that they can be localized at the interface resulting in effective compatibilization. Janus nanomaterials are an exciting class of compatibilizers as they can be strategically localized at the interface of the immiscible polymer blend system. Their synthesis and application in immiscible polymers are of great significant importance in stabilizing and imparting functionality in immiscible polymer blend systems. The major drawback of Janus nanomaterials is the difficulty in synthesizing them. This means that nanoparticles are commonly used in compatibilization because they are readily available and easy to synthesize. In general, the compatibilization efficiency of nanomaterials is mainly due to the restriction of coalescence mainly because of the polymer–nanomaterial interaction, an increase of viscosity of one of the phases, and the interaction of the nanomaterials with both polymers, especially when lodged at the interphase. The localization of the nanomaterials also plays a significant role in the compatibilization process. Improvement in the interfacial adhesion of two polymers results in better mechanical properties, and these multiphasic structures can be used in different applications.

References

[1] Muzata, T. S., Jagadeshvaran, P. L., Bose, S. Nanoparticles influence miscibility in LCST polymer blends: From fundamental perspective to current applications. *Phys. Chem. Chem. Phys.* 22, 20167–20188 (2020)

[2] Gebrekrstos, A., Orasugh, J. T., Muzata, T. S., Ray, S. S. Cellulose-based sustainable composites: a review of systems for applications in EMI shielding and sensors. *Macromol. Mater. Eng.* 307, 2200185 (2022)

[3] Xavier, P., Jain, S., Srinivas, V., Chatterjee, K., Bose, S. Designer porous antibacterial membranes derived from thermally induced phase separation of PS/PVME blends decorated with an electrospun nanofiber scaffold. *RSC Adv.* 6, 10865–10872 (2016)

[4] Muzata, T.S., Bose, S. Polymer tethered graphene oxide influences miscibility and cooperative relaxation in LCST blends. *Polymer* 188, (2020)

[5] Muzata, T.S., Jagadeshvaran, P.L., Kar, G.P., Bose, S. Phase miscibility and dynamic heterogeneity in PMMA/SAN blends through solvent free reactive grafting of SAN on graphene oxide. *Phys. Chem. Chem. Phys.* 20, (2018)

[6] Sharma, M., Sharma, K., Bose, S. Segmental relaxations and crystallization-induced phase separation in PVDF/PMMA blends in the presence of surface-functionalized multiwall carbon nanotubes. *J. Phys. Chem. B* 117, 8589–8602 (2013)

[7] Ding, Y. et al. PLA-PEG-PLA tri-block copolymers: Effective compatibilizers for promotion of the interfacial structure and mechanical properties of PLA/PBAT blends. *Polymer* 146, 179–187 (2018)

[8] Zheng, X., Li, Y., Tang, J., Yu, G. Structure and properties of PVDF/PA6 blends compatibilized by ionic liquid-grafted PA6. *ACS Omega* 7, 12772–12778 (2022)

[9] Wang, Y. et al. Insight on compatibilization of LLDPE/PS blends from morphology, interfacial state, mechanical properties and melt properties: Comb-like copolymer vs diblock copolymer. *Polymer* 218, 123540 (2021)

[10] Tamiya, T. et al. Enhancement of interfacial adhesion in immiscible polymer blend by using a graft copolymer synthesized from propargyl-terminated poly(3-hydroxybutyrate-co-3-hydroxyhexanoate). *Eur. Polym. J.* 130, 109662 (2020)

[11] Poggi, E., Gohy, J.F. Janus particles: from synthesis to application. *Colloid Polym. Sci.* 295, 2083–2108 (2017)

[12] Seyni, F.I., Grady, B.P. Janus particles as immiscible polymer blend compatibilizers: a review. *Colloid Polym. Sci.* 299, 585–593 (2021)

[13] Lattuada, M., Hatton, T.A. Synthesis, properties and applications of Janus nanoparticles. *Nano Today* 6, 286–308 (2011)

[14] Xu, W. et al. Study on the compatibilizing effect of Janus particles on liquid isoprene rubber/epoxy resin composite materials. *Ind. Eng. Chem. Res.* 56, 14060–14068 (2017)

[15] You, W., Yu, W. Onset reduction and stabilization of cocontinuous morphology in immiscible polymer blends by snowmanlike Janus nanoparticles. *Langmuir* 34, 11092–11100 (2018)

[16] Walther, A., Matussek, K., Müller, A.H.E. Engineering nanostructured polymer blends with controlled nanoparticle location using Janus particles. *ACS Nano* 2, 1167–1178 (2008)

[17] Parpaite, T. et al. Janus hybrid silica/polymer nanoparticles as effective compatibilizing agents for polystyrene/polyamide-6 melted blends. *Polymer* 90, 34–44 (2016)

[18] Bahrami, R., Löbling, T.I., Schmalz, H., Müller, A.H.E., Altstädt, V. Synergistic effects of Janus particles and triblock terpolymers on toughness of immiscible polymer blends. *Polymer* 109, 229–237 (2017)

[19] Bärwinkel, S. et al. Polymer foams made of immiscible polymer blends compatibilized by Janus particles – Effect of compatibilization on foam morphology. *Adv. Eng. Mater.* 18, 814–825 (2016)

[20] Cao, Y., Zhang, J., Feng, J., Wu, P. Compatibilization of immiscible polymer blends using graphene oxide sheets. *ACS Nano* 5, 5920–5927 (2011)

[21] Tong, J., Huang, H.X., Wu, M. Promoting compatibilization effect of graphene oxide on immiscible PS/PVDF blend via water-assisted mixing extrusion. *Compos. Sci. Technol.* 149, 286–293 (2017)

[22] Ye, S., Cao, Y., Feng, J., Wu, P. Temperature-dependent compatibilizing effect of graphene oxide as a compatibilizer for immiscible polymer blends. *RSC Adv.* 3, 7987–7995 (2013)

[23] Yan, N. et al. Compatibilization of natural rubber/high density polyethylene thermoplastic vulcanizate with graphene oxide through ultrasonically assisted latex mixing. *J. Appl. Polym. Sci.* 127, 933–941 (2013)

[24] Nasrollah Gavgani, J., Goharpey, F., Velankar, S. Interfacially compatibilized PI/PDMS blends with reduced octadecylamine-functionalized graphene oxide: morphological and rheological properties. *Soft Matter* 17, 9670–9681 (2021)

[25] Cao, Y., Feng, J., Wu, P. Polypropylene-grafted graphene oxide sheets as multifunctional compatibilizers for polyolefin-based polymer blends. *J. Mater. Chem.* 22, 14997–15005 (2012)

[26] Taraghi, I., Fereidoon, A., Paszkiewicz, S., Roslaniec, Z. Nanocomposites based on polymer blends: enhanced interfacial interactions in polycarbonate/ethylene-propylene copolymer blends with multi-walled carbon nanotubes. *Compos. Interfaces* 25, 275–286 (2018)

[27] Chen, J. et al. Improving interfacial adhesion between immiscible polymers by carbon nanotubes. *Polymer* 54, 464–471 (2013)

[28] Wu, D., Zhang, Y., Zhang, M., Yu, W. Selective localization of multiwalled carbon nanotubes in poly(ε-caprolactone)/polylactide blend. *Biomacromolecules* 10, 417–424 (2009)

[29] Zhao, X., Wang, H., Fu, Z., Li, Y. Enhanced interfacial adhesion by reactive carbon nanotubes: new route to high-performance immiscible polymer blend nanocomposites with simultaneously enhanced toughness, tensile strength, and electrical conductivity. *ACS Appl. Mater. Interfaces* 10, 8411–8416 (2018)

[30] Sharma, M., Madras, G., Bose, S. Cooperativity and structural relaxations in PVDF/PMMA blends in the presence of MWNTs: An assessment through SAXS and dielectric spectroscopy. *Macromolecules* 47, 1392–1402 (2014)

[31] Sharma, M., Madras, G., Bose, S. Unusual fragility and cooperativity in glass-forming and crystalline PVDF/PMMA blends in the presence of multiwall carbon nanotubes. *Macromolecules* 48, 2740–2750 (2015)

[32] Liu, H. et al. Design of high-performance poly(L-lactide)/elastomer blends through anchoring carbon nanotubes at the interface with the aid of stereocomplex crystallization. *Polymer* 108, 38–49 (2017)

9 Applications of Immiscible Polymer Blends

9.1 Introduction

Even though polymer composites are widely being used in different applications [1], the advent of polymer blends has also found great use in different applications [2–5]. Recent environmental problems, such as water and air pollution and the presence of unwanted irradiation in the environment, have resulted in increased environmental peril. This might be due to rapid industrialization which is currently taking place all over the world. Hence, this requires robust and effective mechanisms to prevent such pollution, which unattended might have catastrophic effects. The rapid increase in the number of people living on this earth has also led to a demand for freshwater from rivers, lakes, etc. To curb the demand for freshwater, new technologies must be implemented to utilize unused water sources. Several technologies have been implanted for water treatment, and those include chemical methods, physical methods, which include filtration, and biological methods, which make use of anaerobic reactors. Membrane technology is one of the most effective ways used in water remediation due to its numerous advantages, such as excellent separation efficiencies and low cost of operation. Membranes made of polymeric materials can allow only water molecules to pass through them and exclude the passage of other solutes due to size. Polymeric membranes are mainly used for desalination and waste treatment [6]. Polymeric materials can be used to fabricate systems that are capable of purifying water, detecting the levels of noxious gases present in the atmosphere, and also prevent electromagnetic interference [7]. Polymeric materials are now the favored choice of materials because of the low cost of production, tunable properties for specific applications, and good antifouling properties for water remediation applications [8–10]. This chapter will specifically explore the different applications of both compatibilized and noncompatibilized immiscible polymer blend systems.

9.2 Gas and Moisture Barrier Applications

Excellent gas barrier properties are a major requirement that is of paramount importance in different applications, such as food packaging, biomedical packaging, and wound dressing, etc. Polymers are now widely used in packaging applications, mainly food packaging, because they offer several advantages, such as they are lightweight and easy to process. Even though they offer numerous advantages compared to other materials, polymers are usually permeable to small gaseous molecules. The diffusion of gaseous molecules across the polymer film is mainly governed by several factors, such as environmental temperature and humidity, film thickness, and pressure differences. The movement of gaseous molecules across a polymer thin film is mainly via diffusion. The molecules move from a higher concentrated side by dissolving through the thin film and then evaporate from the other side. In an endeavor to improve the barrier properties of polymeric materials, researchers utilized nanomaterials and blending. Multiphasic systems composed of polymeric materials can be used in gas barrier applications. This application is very important to prevent product deterioration and contamination, especially in the packaging industry. Improvement of the gas barrier can be made by blending two polymers, and as stated earlier, blending can lead to miscible and immiscible polymer systems. PP is one of the most used commodity plastics in the world, and it can be blended with polyamide-66 (PA66) to improve its barrier properties. The main disadvantage of these two blends is the dispersion of PA66. This can be improved by forming microlayers of PA66 dispersed in the PP matrix [11]. The orientated microplatelets of PA66 provide a tortuous pathway for gas molecules to pass through. Microlayer coextrusion can be an effective way of fabricating polymer blend systems that have superior barrier properties against gaseous molecules. Boufarguine et al. made use of a multilayer co-extrusion setup to blend poly(lactic acid) (PLA) and poly(3-hydroxybutyrate-co-3-hydroxyvalerate) (PHBV) [12]. PLA is biodegradable and degrades into water and carbon dioxide and it is produced from polysaccharide fermentation. They produced a film with alternating layers of the polymers. PHBV enhanced the barrier properties of the blend compared to neat polymer. The incorporation of nanoparticles significantly improves the movement of gaseous molecules through a polymer blend system. Different types of nanoparticles, such as nanoclay, nano-titanium oxide, graphene, and nanosilica can be used to provide a tortuous pathway for the diffusion of gas molecules. Halloysite nanotubes (HNTs) are clay nanomaterials that can be added to polymer blends to ameliorate their barrier properties, they are hydrophobic due to the siloxane groups and possesses good mechanical properties. These nanomaterials have been used in different polymer composites to improve their barrier properties [13]. Abdullah et al. [14] made use of poly(vinyl alcohol) (PVA)/starch and added HNTs into the ecofriendly polymer blend system. The presence of these nanomaterials enhanced the gas and water barrier properties of the system. These blends find great use in the packaging of foodstuff. The addition of clay nanomaterials prevents, to a greater extent, the movement of gas molecules by decreasing the area in which the molecules diffuse by replacing the permeable polymer portions with an impervious layer and by increasing the distance in which the gaseous molecules travel hence providing a tortuous pathway for them. Frounchi et al. [15] observed an increase in gas obstruction of PP/ethylene-propylene-diene rubber (EPDM) when they added organoclay into the polymer blend system. They further explained that significant gas impediment could be obtained when clay nanomaterials with a high aspect ratio are used. These flakes provide a tortuous path that slows down the diffusion of the gaseous molecules. Maria et al. [16] also reported that the presence of organoclay also reduced the gas barrier properties of a polymer blend of natural rubber (NR)/nitrile rubber (NBR). They also reported that the diffusion of the gaseous molecules depended upon the blend composition, gaseous nature, and the concentration of organoclay incorporated into the polymer blend. Polymers generated from natural resources are usually biodegradable and favored in the production of packaging materials that have good moisture and gas barrier properties. One of the most common natural polymers is starch. Starch possesses good oxygen barrier properties because of its crystalline nature. The major disadvantage of starch is its poor mechanical and thermal properties. Its properties can be enhanced by mixing it with different polymers. Chang and co-workers blended thermoplastic starch and poly(butylene adipate-co-terephthalate) (PBAT) [17]. To improve the interaction between the two polymers, they used maleic anhydride, and the polymer blends registered an enhancement in moisture and oxygen barrier properties. Addition of the hydrophobic PBAT (which was used to coat and blend) improved the water barrier properties of the thermoplastic starch, because starch is a hydrophilic polymer hence its water barrier properties are very poor as shown in Figure 9.1.

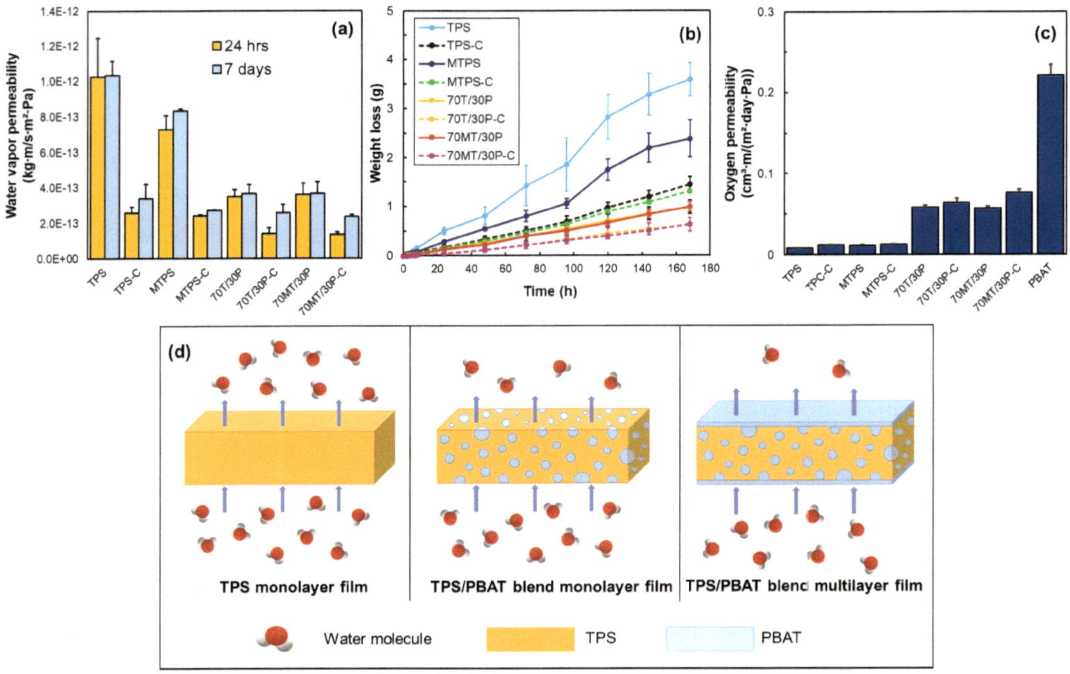

Figure 9.1 (a) Water vapor permeability of films over 24 h and 7 days; (b) water vapor loss over times; (c) oxygen permeability of films; (d) illustration of water vapor permeation mechanism for TPS/PBAT monolayer and multilayer films. Error bars show standard deviation from three replicates. Reprinted with permission from [17]. Copyright 2021, Elsevier Science Ltd

9.3 Gas Sensing Applications

Due to rapid industrialization, many toxic gases are now being emitted into the environment, and these gases are detrimental to the environment, human beings, and animals. It is of paramount importance to detect these gases as early as possible to avoid catastrophic effects. The use of polymers as gas sensors has multiple advantages over metal because they are lightweight and inexpensive, they can be used at low temperatures, and they are easy to fabricate [18]. Polymer blends filled with conducting nanomaterials can be used to fabricate sensors [19]. The conducting networks formed by the nanomaterials can change their electrical properties when subjected to external stimuli. The changes in this conducting network may be due to some changes in the polymer structure; for example, the polymer may swell when it is subjected to specific vapors. The development of polymer-based sensors from this concept is very important for detecting leakages. CNTs and graphene nanoparticles are some of the nanomaterials that can be incorporated into polymer blend systems to make them conduct. Li et al. [20] prepared a PC/PS polymer blend system and incorporated MWCNT to induce conductivity in the system for volatile organic compound sensing applications. The two polymers constituting the polymer blend were selected such that they had different interactions with the organic vapors under test. Immiscible polymer blend systems containing carbon nanotubes can be used as vapor sensors. Also, interestingly, vapor sensing can be used to ascertain the localization of carbon nanotubes in immiscible polymer blends [21]. The polymer and vapor interaction is mainly governed by the Flory–Huggins theory. A low value of the Flory–Huggins interaction parameter shows good interaction of the polymer and vapor. There are other gases, such as ammonia that must be detected early because of their capabilities to cause lung-associated diseases. Ammonia is a toxic gas that is usually produced in the chemical industry and from vehicles. Polymeric materials always find great use in detecting gases such as ammonia because they are inexpensive, as mentioned earlier, compared to other spectroscopy methods. Conducting polymers blended with an insulating polymer are now being used to detect gas leakages due to multiple advantages. The presence of the π-electrons in them enables them to have good sensitivity. Organic field-effect transistors (OFETs) are used in detecting ammonia gas, and Han et al. [22] made use of poly(3-hexylthiophene) (P3HT) and PS to make a polymer blend that was capable of good air stability and also having high mobility ratios. 6,13-Bis[(triisopropylsilyl)ethynyl]pentacene TIPS-pentacene/PS blend films were used as a gas sensor when exposed to nitrogen dioxide gas (NO_2); the mechanism in which the blends were able to sense was mainly as a result of the dipolar nature of NO_2 [23]. The NO_2 gas causes the accumulation of extra hole carriers when situated between the interface of PS and TIPS-pentacene, increasing hole carrier density. Thin, porous structures have been reported to allow NO_2 to easily move through them, hence increasing sensitivity. Figure 9.2 shows the sensing mechanism of the polymer blend system.

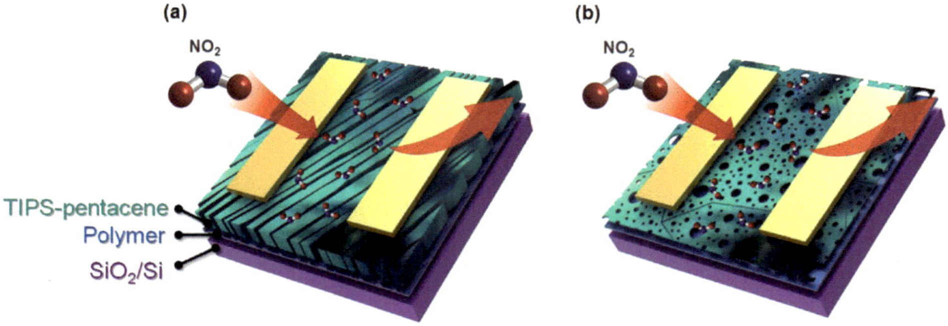

Figure 9.2 Schematic diagram showing gas sensing mechanism of TIPS-pentacene/PS FETs prepared with different spinning times: (a) 5 s, (b) 50 s. Reproduced from [23].

The combination of a conducting and insulating polymer proved to be an effective way of developing portable and inexpensive ammonia gas sensors. Another conducting polymer that can be used in gas sensing is PANI (polyaniline). PANI has found great use in the fabrication of sensors mainly due to its excellent stability under ambient conditions and its ability to change their conductivity during potential switching. The processability of PANI can be enhanced by mixing it with different types of polymers that are easy to process, such as PS. PANI can be blended with another different type of polymer, such as PI, to increase the overall mechanical properties without significantly altering its electrical conductivity. The addition of PI to PANI was shown to improve the sensitivity of PANI toward carbon monoxide (CO) [24]. Polypyrrole (PPy) is another conducting polymer that can be utilized in the fabrication of materials that can be used in sensing applications. PPy can be blended with other polymers, such as phenothiazine, to make a blend that can be used in detecting ammonia gas [25]. The incorporation of conducting nanoparticles also enhances the sensing properties of the blends. Ramesan et al. [25] added silver-zinc oxide (Ag-ZnO) nanomaterial to improve the overall electrical conductivity of the system and also its sensing properties against ammonia gas. The sensing performance of a conducting polymer is mainly based on the chemical interaction between the sensor material and the gas molecules, which changes the doping state of the conducting polymers. This simple technique allows the conducting polymer materials to be used as gas sensors.

■ 9.4 Membranes for Water Remediation

Due to several factors, such as rapid population growth and global warming, water demand has significantly increased. It is of great importance to use water from current sources and recycle waste water. Porous membranes have now been widely utilized to develop membranes for water remediation. Polymer blend systems provide are one of the best templates in water remediation because they can be tuned to create favorable morphologies, either co-continuous or droplet matrix morphologies. By tailoring the interface of two polymers, Mural et al. [26] developed porous membranes of polyethylene (LDPE)/poly(ethylene oxide) (PEO). Different types of compatibilizers, such as graphene oxide (GO), polyethylene grafted GO (PE-g-GO), and amine-terminated GO (GO–NH_2), were used. The interfacial adhesion between the two immiscible polymers was noticed and significantly improved with the addition of GO–NH_2 with maleated PE. The presence of GO–NH_2 significantly improved the antibacterial properties of the membrane. Using the same polymer blend system, Mural et al. [27] developed another antibacterial membrane of different types of nanoparticles, such as silver (Ag) nanoparticles, reduced graphene oxide (rGO), and a combination of both Ag and rGO (rGO–Ag). A composition of 90/10 was selected, and the water-soluble phase of PEO was etched out to fabricate a porous template. The addition of nanoparticles was shown to reduce the flux performance, and this was due to the reduction in pore size. Both nanoparticles imparted antibacterial properties on the membrane. Mural et al. [28] also imparted antibacterial properties on the similar polymer blend system LDPE/PEO by grafting poly(ethylene imine) (PEI) and depositing Ag nanoparticles on the membrane. They controlled the blend morphology by making use of a compatibilizer, maleated PE. The polymer blend system, which contained the compatibilizer, was shown to have refined morphology, which is of great significance because this stabilizes the system during industrial processes that require elevated temperatures. The polymer blend system containing grafted PEI and Ag nanoparticles showed significant antibacterial properties (almost 100% bactericidal efficacy); this resulted from the synergistic effect of PEI and Ag nanoparticles that disrupts the integrity of the bacterial cells and liberation of cell intracellular electrolytes. During the water filtration process, particles such as microbes and other pollutants tend to attach to the surface of the membranes leading to pore blockage. Remanan et al. [29] developed a surface-modified membrane composed of poly(ethylene-co-methyl acrylate) (EMA)/poly(vinylidene fluoride) (PVDF). The surface of the membrane was modified by PEI/rGO to induce antifouling properties. The porous membrane was obtained by selectively etching out the PVDF phase. In another work by Remanan et al. [30], they developed an EMA/PVDF porous membrane that was decorated with both phase-transited lysozyme (PTL) nanoparticles and molybdenum disulfide (MoS_2) nanosheets in an endeavor to impart antifouling and antibacterial properties on the membrane. It was observed from experimental results that in terms of antibacterial properties, the polymer membrane modified with PTL exhibited higher antibacterial properties in comparison to the membranes that were modified with a combination of chitosan and MoS_2. In terms of antifouling properties, the chitosan/MoS_2-modified membrane registered higher antifouling properties than the PTL-modified membrane. This was mainly attributed to the interaction between the lysosome and bovine serum albumin.

■ 9.5 Electromagnetic Interference Shielding Materials

Due to the development of different types of electronic gadgets, the rise of electromagnetic (EM) radiation has led to EM pollution. Different types of EM shielding materials have been developed, but recently researchers have focused on the development and use of polymer blend composites. The blending of two immiscible polymers is an effective way to develop polymeric materials that can shield electromagnetic radiation. This is mainly because polymer blend composites are lightweight, they can be manipulated to alter the properties of the overall blend system, and they do not undergo corrosion [31]. Effective attenuation of electromagnetic radiation is mainly possible if the polymer blend system possesses high electrical conductivity and magnetic loss. EM radiation can be shielded by different mechanisms, such as reflection, multiple reflection, and absorption. The reflection mechanism usually occurs in metals and other conducting materials, and the absorption mechanism occurs in materials having electric and magnetic properties; in polymers, this can be achieved by incorporating nanomaterials possessing those properties. The incorporation of multiwall carbon nanotubes (MWNTs) imparts electrical conductivity in the polymer blend system. MWNT are an important class of nanoparticles because they are conducting in nature; they have a high aspect ratio enabling them to facilitate the formation of an interconnected 3D structure. MWCNT are made from multiple rolls of graphene sheets with an interlayer spacing of approximately 0.34 nm and an outer diameter ranging from 2.5 to 100 nm [32]. Phase separation can be utilized to control the dispersion of nanomaterials such as CNTs in multiphasic systems. This will result in the fabrication of soft conducting templates that can be used to develop EMI shielding materials. Bose et al. [33] studied how conductivity can be enhanced by phase separation in poly[(α-methylstyrene)-co-acrylonitrile]/poly(methyl methacrylate) in the presence of modified MWCNTs. In a single phase, the polymer blend registered some insulating characteristics, but after phase separation, there was an increase in conductivity. Interestingly, He et al. [34] also used phase separation as a tool in fabricating a template that was used for EMI shielding. They used epoxy (EP) and polyether sulfone (PES) in the presence of graphene and MWCNT. After the blend demixes via reaction-induced phase separation (see Figure 9.3), the MWCNT was found to be located in the EP phase and the graphene in the PES phase. The polymer blend system shows low critical solution temperature (LCST) behavior, usually, these types of blend systems phase separate at high temperatures and are miscible at low temperatures. The strategic localization of conducting nanomaterials enhances the conductivity; as a result improving the EMI shielding ability of the polymer blend composite.

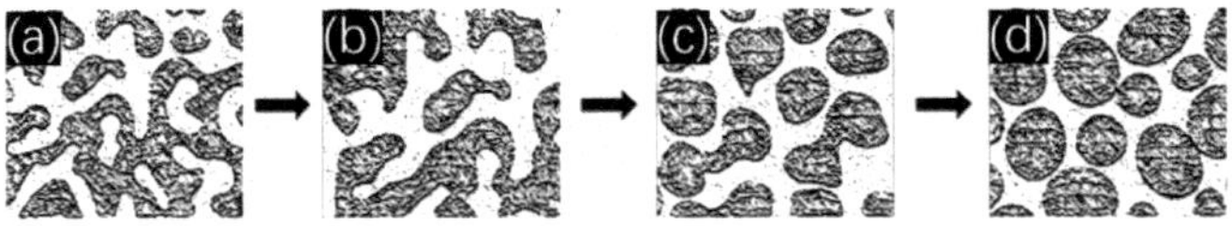

Figure 9.3 Mechanism of reaction-induced phase separation in EP/PES composites. Reprinted with permission from [34]. Copyright 2021, Elsevier Science Ltd

Using similar polymers, Zhang et al. [35] made use of ultrasonic treatment and dichloromethane-assisted dispersion to fabricate a polymer blend composite of EP/PES in the presence of MWCNT. The phase separation process and the localization of the nanomaterials can be seen in Figure 9.4. The formation of a co-continuous structure with the presence of the conducting nanoparticles improved the EMI shielding ability of the blend system.

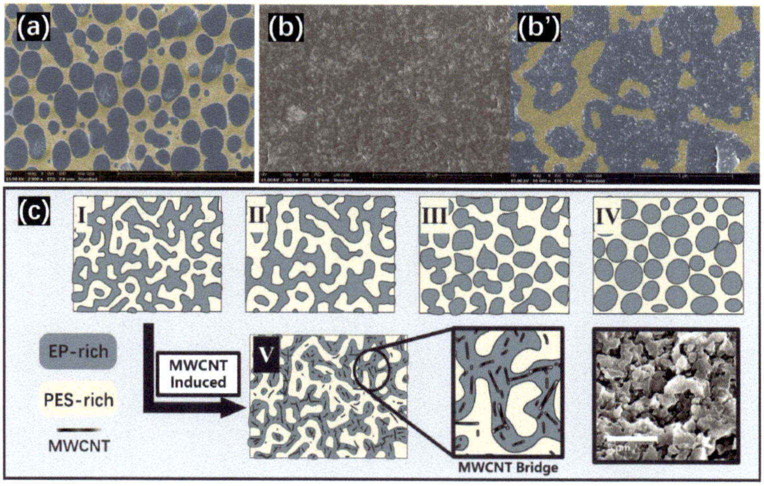

Figure 9.4 Scanning electron microscope images of cryogenically fractured surfaces of (a) EP/PES23 blend and (b, b') EP/PES23/MWCNT2.9 blend after curing at 150 °C for 100 min; (c) schematic reaction-induced phase separation in EP/PES blends: temporary co-continuous structure (I) develops at the early stage of spinodal decomposition and grows self-similarly, resulting in the similar structure with longer periodic distance (II). Then the phase continuity is interrupted by increasing interfacial tension, transforming into a fragmented structure (III) and further into a spherical domain structure (IV); the mechanism of co-continuous structure formation in EP/PES23/MWCNT2.9 is given in (V). Reprinted with permission from [35]. Copyright 2020, Elsevier Science Ltd

Curing reaction-induced phase separation was used by Meng and co-workers [36] to develop a conducting polymer blend system comprising diglycidyl ether of the bisphenol A/polyetherimide (DGEBA/PEI) with reduced graphene oxide (RGO). The conducting RGO was found to be localized at the interface of the blend system. The presence of conducting nanomaterials and their localization on the interface improve the conductivity and the shielding properties of polymer blends. The phase separation process and the shielding mechanism are illustrated in Figure 9.5.

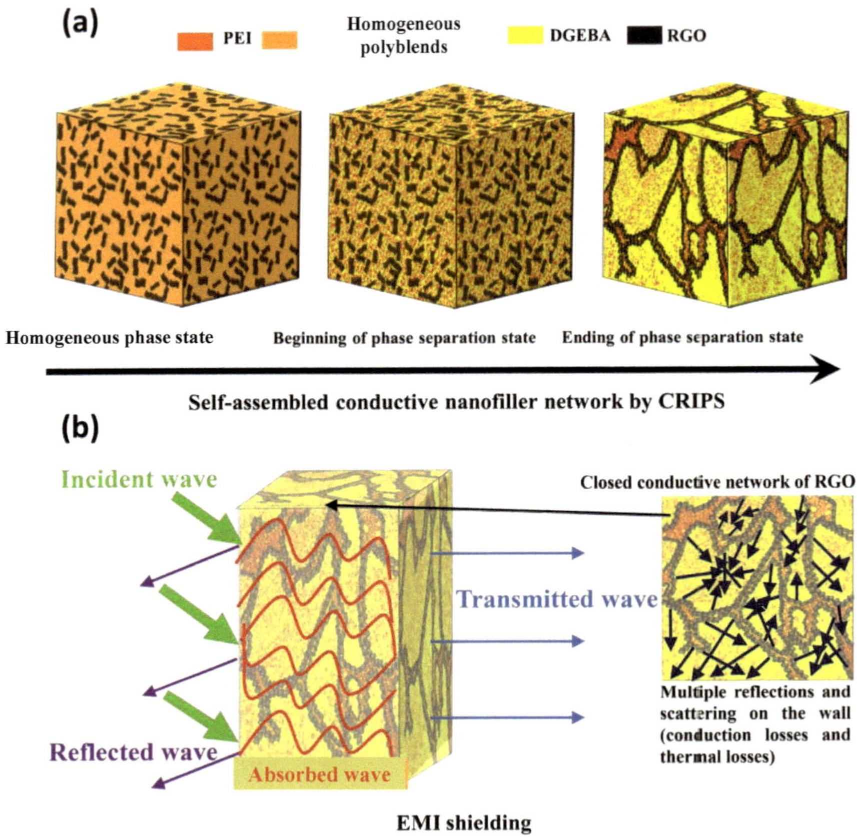

Figure 9.5 Schematic of (a) morphological evolution of double perco ation, conductive DPn1Rn2 nanocomposites and (b) the EMI shielding mechanism of DPn1Rn2 nanocomposites. Reproduced from [36]. It is an open-access publication and was published under a Creative Commons Attribution 3.0 International License.

Rohini et al. [37] fabricated an EMI shielding material by making use of a polymer blend system of PS/PMMA. Different types of MWNTs were incorporated into the polymer system. It was shown that the pristine MWCNT was in the PMMA phase. They also reported that MWNT and amine-functionalized MWNTs (NH_2-MWNT) facilitate interconnected PMMA when the polymer blend system was subjected to annealing (see Figure 9.6). In terms of electrical conductivity, an essential property in EMI shielding materials, the polymer blend systems containing MWNT and NH_2-MWNT domains exhibited high conductivity compared to the neat blend. An increase in conductivity resulted from MWNT gelation in the PMMA phase. The polymer blend system containing carboxylic acid (COOH-MWNTs) functionalized groups registered a marginal increase in conductivity due to less interconnection. These polymer blend systems containing MWNT were proved to be ideal candidates for attenuating EM radiation.

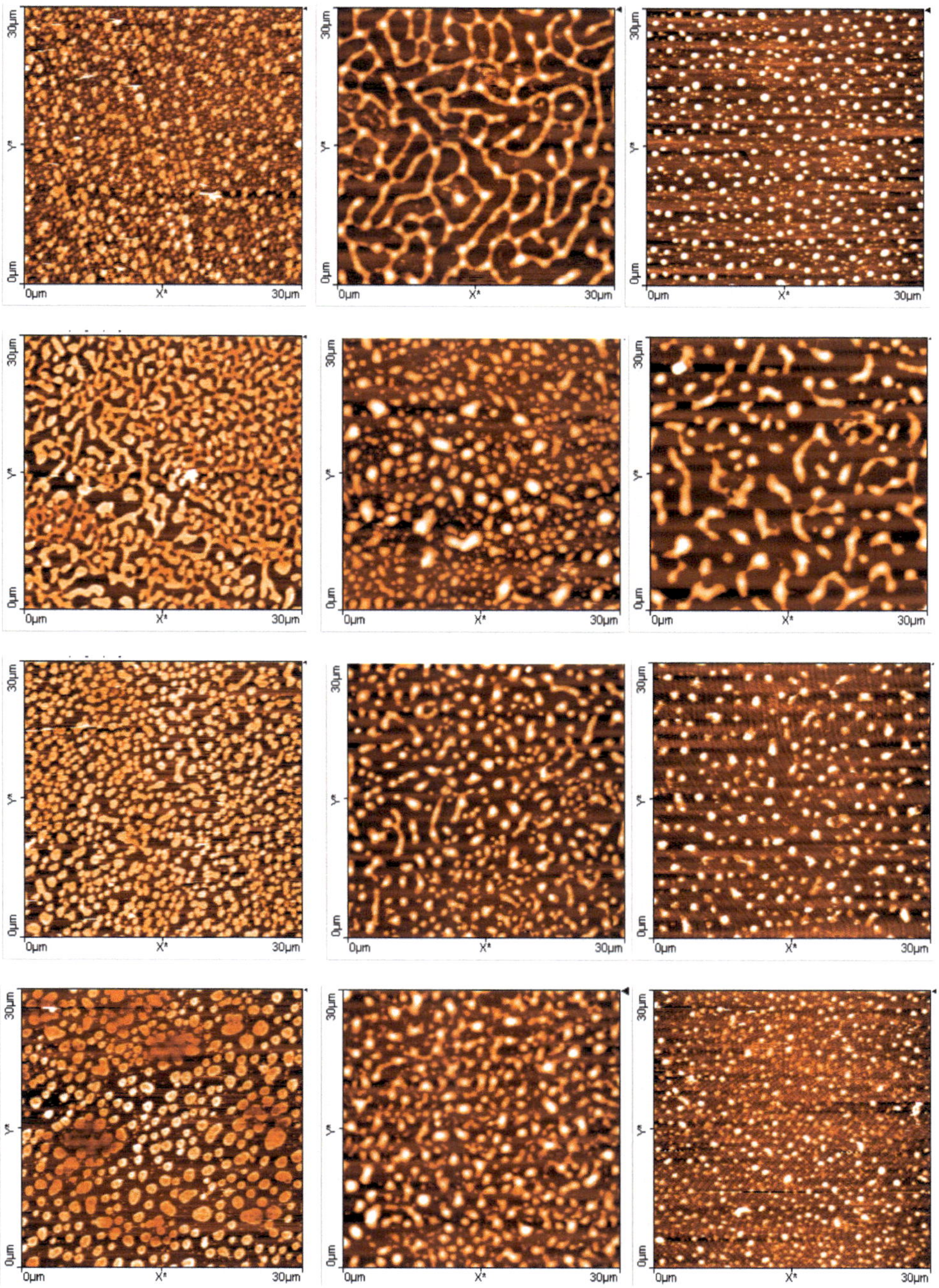

Figure 9.6 Atomic force microscope 2D topographical images of neat blends (1st row), blends with pristine MWNTs (2nd row), NH_2-MWNTs (3rd row), and COOH-MWNTs (4th row) under different conditions. The left column represents stack of images before annealing; the middle column represents after annealing for 30 min; and the right column represents images captured after annealing for 60 min. Reprinted with permission from [37]. Copyright 2014, American Chemical Society

Controlling the nanoparticle localization in an immiscible polymer blend system is an effective way of establishing a material that can effectively attenuate EM radiation. In their study, Biswas et al. [38] incorporated different types of nanomaterials inside a PC/PVDF polymer blend. The nanomaterials used were PANI-wrapped MWNT, barium titanate (BT), and iron(II,III) oxide (Fe_3O_4) [38]. They combined PANI-wrapped MWNT and Fe_3O_4 to produce PANI-MWNT-Fe_3O_4 and BT were terminated with NH_2 groups to produce PC-BT-NH_2 composite. The BT nanomaterial was restricted in the PC phase to prevent the establishment of a conducting network in the PVDF phase since MWNT were found to be in the PVDF phase. This preferential localization of the nanoparticles, which conduct electricity, in a particular phase ensured an increase in conductivity. The presence of BT and Fe_3O_4 is of paramount importance because of their dielectric and magnetic loss properties. The selective localization of MWNT in a specific phase mainly depends on factors such as kinetics and thermodynamics. It is important to ensure that the nanomaterials are locked at the interface to improve the conductivity of the polymer blend system. Kar and co-workers [39] established an interesting approach to driving MWNTs to the interface of a polymer blend system comprising PVDF and acrylonitrile butadiene styrene (ABS). They wrapped MWNT with PMMA, a polymer that interacts mutually with PVDF and ABS. The localization of the PMMA-wrapped MWNT at the interface of the two immiscible polymers causes an interfacial tension reduction. The PMMA-wrapped MWNT registered significant shielding effectiveness which was due to the 3D network structure established at the interface. In polymer blends containing PMMA-wrapped MWNT, the shielding mechanism was mainly due to absorption, which was attributed to the PMMA high dielectric constant. In another different work, Kar and co-workers [3] went on to use different strategies to attenuate EM radiation. They modified MWNT with ionic liquid (IL) to facilitate the formation of a network structure and also improve their dispersion. Agglomeration of nanoparticles inside a polymer matrix can cause a decrease in the electrical properties of the overall composite. It is important to improve the dispersion of nanomaterials to achieve maximum results. The MWNT modified with IL showed a significant increase in electrical conductivity, mainly because of the network structure formation facilitated by the presence of IL. Electrical conductivity is an important requirement for fabricating EMI shielding materials. They went on also to use GO which was chemically attached to barium titanate (BT). The BT-GO, together with IL-MWNT, improved the attenuation performance through the absorption of the overall blend system. It is important to carefully consider the type of CNTs added into the demixed blend system. Different types of CNTs can yield different electromagnetic shielding performances [40]. MWCNT were incorporated into the ethylene methyl acrylate (EMA)/ethylene octane copolymer (EOC) polymer blend. The presence of MWCNT improved the shielding performance of the blend system due to the double percolation phenomena [41]. The shielding was due to multiple reflections, absorption, and reflection, as shown in schematically in Figure 9.7.

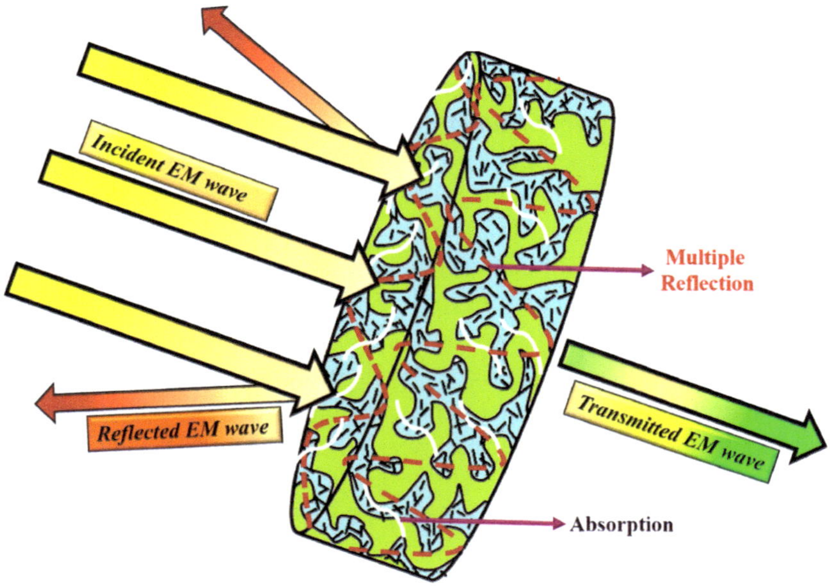

Figure 9.7 Schematic representation of EMI shielding of MWCNT filled binary polymer composite. Reprinted with permission from [41]. Copyright 2019, Elsevier Science Ltd

Another nanomaterial that can be added into immiscible blends for EMI shielding is carbon black (CB). CB is a material with good stability; it is also conducting and has a high surface area. CB was incorporated into an immiscible polymer blend system of ultrahigh-molecular weight PE (UHMWPE)/PP [42]. The polymer blend composites were able to reflect the EM radiation because of the interaction between free electrons, and the EM radiation and absorption were due to the interaction of the electromagnetic fields with the electric dipoles in the system. Rani and co-workers [43] incorporated CB into a polymer blend system of PVA/PEG and studied the ability of the polymer blend composite to shield EMI radiation. The increase in the shielding effectiveness was due to the conductive pathway created by the CB in the system. Ensuring double percolation in an immiscible polymer blend system is one way of enhancing the EMI shielding of carbon black containing polymer blend composites. Ravindren et al. [44] made use of an ethylene methyl acrylate (EMA)/ethylene octane copolymer (EOC) blend system in the presence of CB. The nanoparticles were localized in the EMA phase of the system due to the polar-polar interaction between the polymer and the nanoparticles. The presence of the nanomaterials in the EMA phase resulted in double percolation, which improved the EMI shielding performance of the blend composite. Soares et al. [45] reported good shielding effectiveness of CB in PS/EVA (ethylene vinyl acetate copolymer) blend. The CB was localized in the EVA phase, and the co-continuous morphology was ideal for double percolation. Graphene nanoplatelets (GNP) are a carbon derivative; they are flat sheets having a 2D honeycomb structure. They possess high electrical conductivity, which is one of the major requirements in attenuating EMI radiation when incorporated into a polymer blend. The mechanism of attenuation in polymer composites having conducting nanoparticles is mainly governed by the amount of nanomaterial incorporated into the polymer blend system. Pawar et al. [46] decorated graphene sheets with nickel nanoparticles (G-Ni) and incorporated them into an immiscible polymer blend of polycarbonate/poly(styrene-co-acrylonitrile) (PC/SAN). The (G-Ni) nanomaterials were found to be localized in the PC phase. The presence of the G-Ni significantly improved the shielding effectiveness of the polymer blend system compared to those which only had graphene sheets. This was because G-Ni nanomaterials were evenly distributed throughout the system. The presence of the G-Ni resulted in 99.9% attenuation of the EM radiation. MXenes are some of the fillers that can be incorporated into a polymer matrix for EMI shielding applications. Enhanced shielding performance of MXenes can be obtained by modifying with other nanomaterials, such as graphene oxide, as shown in Figure 9.8. To counteract the poor shielding performance of double percolated systems, Wang et al. [47] made use of MXenes in a PS/PVDF system. The shielding mechanism of the blend system was attributed to reflection, absorption, and internal multiple reflections of radiations.

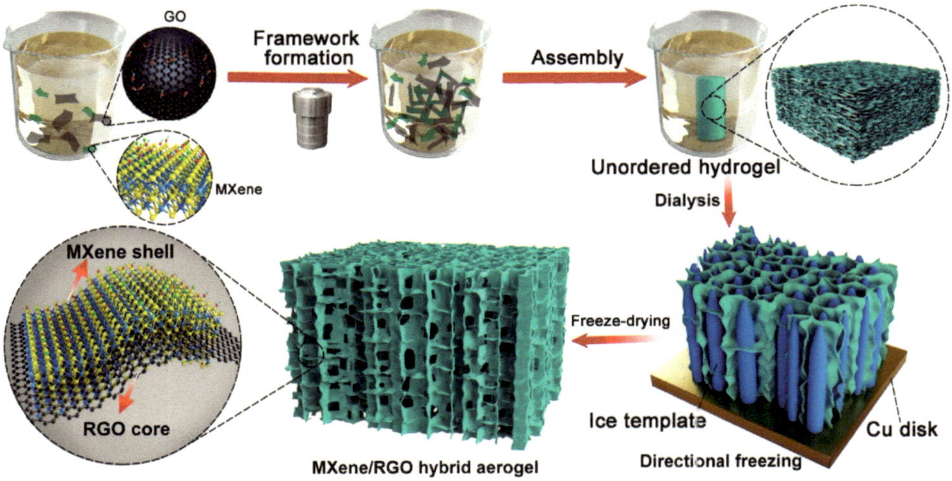

Figure 9.8 Schematic illustrating the fabrication process of a $Ti_3C_2T_x$ MXene/RGO hybrid aerogel by GO-assisted hydrothermal assembly, directional freezing, and freeze-drying Reprinted with permission from [48]. Copyright 2018, American Chemical Society

■ 9.6 Conclusions

The advent of polymer blends has proven to be significant in several applications. Polymer blends can be used in barrier properties, especially in food packaging applications where they extend the shelf life of different food products due to the inhibition of gas and moisture. The application of polymer blends has also been extended to gas sensing. Recently, nonpolymeric materials have been common in gas sensing applications, but the emergence of polymeric materials has seen the advent of sensors being fabricated from polymers. Both conducting polymers and nonconducting polymers have been used in the fabrication of gas sensors, and phase separation has been seen to act as a template in the distribution of conducting nanomaterials at the interface of one of the phases improving the sensing properties of the polymer blend systems. Immiscible polymer blends have also been used in the fabrication of porous membranes, which are used in water remediation. One of the phases was etched out to establish pores that facilitate the transportation of water molecules and hinder other solutes from passing through. The presence of the nanomaterials in the polymer blend system not only improved the compatibility of the two polymers, but also acted as antimicrobial materials that were capable of inactivating different microbes found in water sources. Electromagnetic shielding was observed to be made possible by the emergence of polymer blends. Different types of polymer blend systems with conducting and magnetic nanomaterials can be tailored to establish materials that can shield electromagnetic radiation.

References

[1] Kumar, N., Mahale, B., Muzata, T.S., Ranjan, R. Energy harvesting with flexible piezocomposite fabricated from a biodegradable polymer. *Int. J. Energy Res.* 45, 19395–19404 (2021)

[2] Muzata, T.S., Jagadeshvaran, P.L., Bose, S. Nanoparticles influence miscibility in LCST polymer blends: From fundamental perspective to current applications. *Phys. Chem. Chem. Phys.* 22, 20167–20188 (2020)

[3] Kar, G.P., Biswas, S., Rohini, R., Bose, S. Tailoring the dispersion of multiwall carbon nanotubes in co-continuous PVDF/ABS blends to design materials with enhanced electromagnetic interference shielding. *J. Mater. Chem. A* 3, 7974–7985 (2015)

[4] Xavier, P., Jain, S., Srinivas T, V., Chatterjee, K., Bose, S. Designer porous antibacterial membranes derived from thermally induced phase separation of PS/PVME blends decorated with an electrospun nanofiber scaffold. *RSC Adv.* 6, 10865–10872 (2016)

[5] Pawar, S.P., Bose, S. Peculiar morphological transitions induced by nanoparticles in polymeric blends: retarded relaxation or altered interfacial tension? *Phys. Chem. Chem. Phys.* 17, 14470–14478 (2015)

[6] Gontarek-Castro, E. et al. Characterization of PVDF/graphene nanocomposite membranes for water desalination with enhanced antifungal activity. *Water (Switzerland)* 13, 1279 (2021)

[7] Gebrekrstos, A., Orasugh, J.T., Muzata, T.S., Ray, S.S. Cellulose-based sustainable composites: A review of systems for applications in EMI shielding and sensors. *Macromol. Mater. Eng.* 307, 2200185 (2022)

[8] Cheng, X. et al. Finely tailored pore structure of polyamide nanofiltration membranes for highly-efficient application in water treatment. *Chem. Eng. J.* 417, 127976 (2021)

[9] Kumar, P., Madras, G., Bose, S. PVDF/PBSA membranes with strongly coupled phosphonium derivatives and graphene oxide on the surface towards antibacterial and antifouling activities. *J. Membr. Sci.* 548, 203–214 (2018)

[10] Liu, T. et al. Construction of a composite microporous polyethylene membrane with enhanced fouling resistance for water treatment. *J. Membr. Sci.* 618, 118679 (2021)

[11] Jarus, D., Hiltner, A., Baer, E. Barrier properties of polypropylene/polyamide blends produced by microlayer coextrusion. *Polymer* 43, 2401–2408 (2002)

[12] Boufarguine, M., Guinault, A., Miquelard-Garnier, G., Sollogoub, C. PLA/PHBV films with improved mechanical and gas barrier properties. *Macromol. Mater. Eng.* 298, 1065–1073 (2013)

[13] Tas, C. E., Hendessi, S., Baysal, M., Unal, S., Cebeci, F. C., Menceloglu, Y. Z., Unal, H. Halloysite nanotubes/polyethylene nanocomposites for active food packaging materials with ethylene scavenging and gas barrier properties. *Food Bioprocess Technol.* 10, 789–798 (2017)

[14] Abdullah, Z. W., Dong, Y., Han, N., Liu, S. Water and gas barrier properties of polyvinyl alcohol (PVA)/starch (ST)/glycerol (GL)/halloysite nanotube (HNT) bionanocomposite films: Experimental characterisation and modelling approach. *Composites, Part B* 174, 107033 (2019)

[15] Frounchi, M., Dadbin, S., Salehpour, Z., Noferesti, M. Gas barrier properties of PP/EPDM blend nanocomposites. *J. Membr. Sci.* 282, 142–148 (2006)

[16] Maria, H. J. et al. Gas barrier, rheological and mechanical properties of immiscible natural rubber/acrylonitrile butadiene rubber/organoclay (NR/NBR/Organoclay) blend nanocomposites. *Materials (Basel)* 13, 2654 (2020)

[17] Chang, C. C., Trinh, B. M., Mekonnen, T. H. Robust multiphase and multilayer starch/polymer (TPS/PBAT) film with simultaneous oxygen/moisture barrier properties. *J. Colloid Interface Sci.* 593, 290–303 (2021)

[18] Mir, M. A. et al. Utilization of zeolite/polymer composites for gas sensing: A review. Sensors *Actuators, B* 242, 1007–1020 (2017)

[19] Srivastava, S., Tchoudakov, R., Narkis, M. Preliminary investigation of conductive immiscible polymer blends as sensor materials. *Polym. Eng. Sci.* 40, 1522–1528 (2000)

[20] Li, Y., Pionteck, J., Pötschke, P., Voit, B. Organic vapor sensing behavior of polycarbonate/polystyrene/multi-walled carbon nanotube blend composites with different microstructures. *Mater. Des.* 179, 107897 (2019)

[21] Li, Y. et al. Vapor sensing performance as a diagnosis probe to estimate the distribution of multi-walled carbon nanotubes in poly(lactic acid)/polypropylene conductive composites. *Sensors Actuators, B* 255, 2809–2819 (2018)

[22] Han, S. et al. Poly(3-hexylthiophene)/polystyrene (P3HT/PS) blends based organic field-effect transistor ammonia gas sensor. *Sensors Actuators, B* 225, 10–15 (2016)

[23] Lee, J. H. et al. Effect of crystallization modes in TIPS-pentacene/insulating polymer blends on the gas sensing properties of organic field-effect transistors. *Sci. Rep.* 9, 1–9 (2019)

[24] Watcharaphalakorn, S., Ruangchuay, L., Chotpattananont, D., Sirivat, A., Schwank, J. Polyaniline/polyimide blends as gas sensors and electrical conductivity response to CO-N_2 mixtures. *Polym. Int.* 54, 1126–1133 (2005)

[25] Ramesan, M. T., Greeshma, K. P., Parvathi, K., Anilkumar, T. Structural, electrical, thermal, and gas sensing properties of new conductive blend nanocomposites based on polypyrrole/phenothiazine/silver-doped zinc oxide. *J. Vinyl Addit. Technol.* 26, 187–195 (2020)

[26] Mural, P. K. S., Banerjee, A., Rana, S., Shukla, A. Polyolefin based antibacterial membranes derived from PE/PEO blends compatibilized with amine terminated graphene oxide and maleated PE. *J. Mater. Chem. A* 2, 17635–17648 (2014)

[27] Mural, P.K.S. et al. Porous membranes designed from bi-phasic polymeric blends containing silver decorated reduced graphene oxide synthesized via a facile. *RSC Adv.* 5, 32441–32451 (2015)

[28] Mural, P.K.S., Jain, S., Madras, G., Bose, S. Antibacterial membranes for water remediation with controlled leaching of biocidal silver aided by prior grafting of poly(ethylene imine) onto ozone-treated polyethylene. *ChemistrySelect* 2, 624–631 (2017)

[29] Remanan, S., Das, N. C. A unique microfiltration membrane derived from the poly(ethylene-co-methyl acrylate)/poly(vinylidene fluoride) (EMA/PVDF) biphasic blends and surface modification for antifouling application. *Polym. Test.* 79, 106031 (2019)

[30] Remanan, S., Samantaray, P.K., Bose, S., Das, N.C. Phase transited lysozyme particles and MoS2 nanosheets modified elastomer-like antibacterial and antifouling microfiltration membrane derived from poly(ethylene-co-methyl acrylate)/poly(vinylidene fluoride) (EMA/PVDF) blend for water purification application. *Microporous Mesoporous Mater.* 316, 110945 (2021)

[31] Biswas, S., Muzata, T.S., Krause, B., Rzeczkowski, P., Pötschke, P. Does the type of polymer and carbon nanotube structure control the electromagnetic shielding in melt-mixed polymer nanocomposites? *J Compos. Sci.* 4, 9 (2020)

[32] *Design and Applications of Nanostructured Polymer Blends and Nanocomposite Systems.* Thomas, S., Shanks, R., Chandrasekharakurup, S. (Eds.), Elsevier, Oxford (2015)

[33] Bose, S. et al. Phase separation as a tool to control dispersion of multiwall carbon nanotubes in polymeric blends. *ACS Appl. Mater. Interfaces* 2, 800–807 (2010)

[34] He, J., Zhang, H., Chen, Y., Zou, H., Liang, M. Bi-continuous conductive network induced by in-situ phase separation in epoxy composites with enhanced electromagnetic interference shielding performance. *React. Funct. Polym.* 164, 104918 (2021)

[35] Zhang, H. et al. In-situ co-continuous conductive network induced by carbon nanotubes in epoxy composites with enhanced electromagnetic interference shielding performance. *Chem. Eng. J.* 398, 125559 (2020)

[36] Meng, Y. et al. Enhanced electromagnetic interference shielding properties of immiscible polyblends with selective localization of reduced graphene oxide networks. *Polymers (Basel)* 14, 967 (2022)

[37] Rohini, R.B., Bose, S. Electromagnetic interference shielding materials derived from gelation of multiwall carbon nanotubes in polystyrene/poly(methyl methacrylate) blends. *ACS Appl. Mater. Interfaces* 2014, 6, 6, 11302–11310 (2016)

[38] Biswas, S., Kar, G.P., Bose, S. Attenuating microwave radiation by absorption through controlled nanoparticle localization in PC/PVDF blends. *Phys. Chem. Chem. Phys.* 17, 27698–27712 (2015)

[39] Kar, G.P., Biswas, S., Bose, S. Simultaneous enhancement in mechanical strength, electrical conductivity, and electromagnetic shielding properties in PVDF-ABS blends containing PMMA wrapped multiwall carbon nanotubes. *Phys. Chem. Chem. Phys.* 17, 14856–14865 (2015)

[40] Salehiyan, R., Nofar, M., Ray, S.S., Ojijo, V. Kinetically controlled localization of carbon nanotubes in polylactide/poly(vinylidene fluoride) blend nanocomposites and their influence on electromagnetic interference shielding, electrical conductivity, and rheological properties. *J. Phys. Chem. C* 123, 19195–19207 (2019)

[41] Ravindren, R., Mondal, S., Nath, K., Das, N.C. Investigation of electrical conductivity and electromagnetic interference shielding effectiveness of preferentially distributed conductive filler in highly flexible polymer blends nanocomposites. *Composites, Part A* 118, 75–89 (2019)

[42] Cheng, H. et al. Enhancement of electromagnetic interference shielding performance and wear resistance of the UHMWPE/PP blend by constructing a segregated hybrid conductive carbon black-polymer network. *ACS Omega* 6, 15078–15088 (2021)

[43] Rani, P., Ahamed, M.B., Deshmukh, K. Dielectric and electromagnetic interference shielding properties of carbon black nanoparticles reinforced PVA/PEG blend nanocomposite films. *Mater. Res. Express* 7, 064008 (2020)

[44] Ravindren, R., Mondal, S., Bhawal, P., Ali, S. M.N., Das, N. C. Superior electromagnetic interference shielding effectiveness and low percolation threshold through the preferential distribution of carbon black in the highly flexible polymer blend composites. *Polym. Compos.* 40, 1404–1418 (2019)

[45] Soares, B. G., Touchaleaume, F., Calheiros, L. F., Barra, G. M.O. Effect of double percolation on the electrical properties and electromagnetic interference shielding effectiveness of carbon-black-loaded polystyrene/ethylene vinyl acetate copolymer blends. *J. Appl. Polym. Sci.* 133, 1–10 (2016)

[46] Pawar, S. P., Stephen, S., Bose, S., Mittal, V. Tailored electrical conductivity, electromagnetic shielding and thermal transport in polymeric blends with graphene sheets decorated with nickel nanoparticles. *Phys. Chem. Chem. Phys.* 17, 14922–14930 (2015)

[47] Wang, J., Yang, K., Wang, H., Li, H. A new strategy for high-performance electromagnetic interference shielding by designing a layered double-percolated structure in PS/PVDF/MXene composites. *Eur. Polym. J.* 151, 110450 (2021)

[48] Zhao, S. et al. Highly electrically conductive three-dimensional $Ti_3C_2T_x$ MXene/reduced graphene oxide hybrid aerogels with excellent electromagnetic interference shielding performances. *ACS Nano* 12, 11193–11202 (2018)

Index